ELEMENTS OF
NUMBER
THEORY

I. M. VINOGRADOV

Translated from the Fifth Revised Edition by
SAUL KRAVETZ

DOVER PUBLICATIONS, INC.
Mineola, New York

CONTENTS

Chapter I

DIVISIBILITY THEORY

Chapter II

IMPORTANT NUMBER-THEORETICAL FUNCTIONS

Chapter III

CONGRUENCES

iii

Chapter IV

CONGRUENCES IN ONE UNKNOWN

Chapter V

CONGRUENCES OF SECOND DEGREE

Chapter VI

PRIMITIVE ROOTS AND INDICES

SOLUTIONS OF THE PROBLEMS

ANSWERS TO THE NUMERICAL EXERCISES

PRINTER'S NOTE

Throughout this book the a which takes the following form in the formula (a) assumes the following form ($_a$) whenever it is used in the smallest type size, as a subscript or a superscript. Both forms are meant to be the same.

PREFACE

A series of Russian mathematicians—Chebyshev, Korkin, Zolotaryov, Markov, Voronoi and others—have worked on the theory of numbers. One can become acquainted with the content of the classical work of these notable mathematicians in B. N. Delone's book "The Petersburg School of the Theory of Numbers" ("Peterburgskaya shkola teorii chisel," in Russian, 1947).

Soviet mathematicians, working in the field of number theory, have continued the great tradition of their predecessors and have created powerful new methods which have been used to obtain a series of first-class results; in the number theory section of the book "Mathematics in the USSR after 30 years" ("Matematika v SSSR za 30 let," in Russian, 1948) one can find a report on the attainments of Soviet mathematicians in the field of number theory, and the corresponding bibliographical references.

In my book I present a systematic exposition of the fundamentals of number theory within the scope of a university course. A large collection of problems introduces the reader to some of the new ideas in number theory.

This fifth edition of my book differs considerably from the fourth. A series of changes, allowing a simpler exposition, have been made in all the chapters of the book. The most important changes are the merging of the old chapters IV and V into one chapter IV (reducing the number of chapters to six) and the new, simpler proof of the existence of primitive roots.

The problems at the end of each chapter have been essentially revised. The order of the problems is now in complete correspondence with the order of the presentation of the theoretical material. Some new problems have been added; but the number of numbered problems has been substantially reduced. This was accomplished by the unification, under the letters a, b, c, ..., of previously separate problems which were related by the method of solution or by content. All the solutions of the problems have been reviewed; in many cases these solutions have been simplified or replaced by better ones. Particularly essential changes have been made in the solutions of the problems relating to the distribution of n-th power residues and non-residues, and primitive roots, as well as in the estimations of the corresponding trigonometric sums.

I. M. Vinogradov

ELEMENTS OF
NUMBER
THEORY

CHAPTER I

DIVISIBILITY THEORY

§1. *Basic Concepts and Theorems*

a. The theory of numbers is concerned with the study of the properties of integers. By integers we mean not only the numbers of the natural number sequence 1, 2, 3, ... (the positive integers) but also zero and the negative integers: -1, -2, -3, ...

As a rule, in presenting the theoretical material, we will use letters only to denote integers. In the cases in which letters may denote non-integers, if this is not clear in itself, we will mention it specifically.

The sum, difference and product of two integers a and b are also integers, but the quotient resulting from the division of a by b (if b is different from zero) may be an integer or a non-integer.

b. In the case in which the quotient resulting from the division of a by b is an integer, denoting it by q, we have $a = bq$, i.e. *a is equal to the product of b by an integer.* We will then say that a is *divisible by* b or that b *divides* a. Here a is said to be a *multiple* of b and b is said to be a *divisor* of the number a. The fact that b divides a is written as: $b \backslash a$.

We have the following two theorems.

1. *If a is a multiple of m, and m is a multiple of b, then a is a multiple of b.*

Indeed, it follows from $a = a_1 m$, $m = m_1 b$ that $a = a_1 m_1 b$, where $a_1 m_1$ is an integer. But this proves the theorem.

1

2. *If we know that in an equation of the form $k + l + \ldots + n$ $= p + q + \ldots + s$, all terms except one are multiples of b, then this one term is also a multiple of b.*

Indeed, let the exceptional term be k. We have

$$l = l_1 b, \ldots, n = n_1 b, p = p_1 b, q = q_1 b, \ldots, s = s_1 b,$$
$$k = p + q + \ldots + s - l - \ldots - n$$
$$= (p_1 + q_1 + \ldots + s_1 - l_1 - \ldots - n_1)b,$$

proving our theorem.

c. In the general case, which includes the particular case in which a is divisible by b, we have the theorem:

Every integer a is uniquely representable in terms of the positive integer b in the form

$$a = bq + r, \qquad 0 < r < b$$

Indeed, we obtain one such representation of a by taking bq to be equal to the largest multiple of b which does not exceed a. Assuming that we also have $a = bq_1 + r_1$, $0 < r_1 < b$, we find that $0 = b(q - q_1) + r - r_1$, from which it follows (**2, b**) that $r - r_1$ is a multiple of b. But since $|r - r_1| < b$, the latter is only possible if $r - r_1 = 0$, i.e. if $r = r_1$, from which it also follows that $q = q_1$.

The number q is called the *partial quotient* and the number r is called the *remainder* resulting from the division of a by b.

Examples. Let $b = 14$. We have

$$177 = 14 \cdot 12 + 9, \qquad 0 < 9 < 14;$$
$$-64 = 14 \cdot (-5) + 6, \qquad 0 < 6 < 14;$$
$$154 = 14 \cdot 11 + 0, \qquad 0 = 0 < 14.$$

§2. *The Greatest Common Divisor*

a. In what follows we shall consider only the positive divisors of numbers. Every integer which divides all the integers $a, b, \ldots, l$ is said to be a *common divisor* of them. The largest of these common divisors is said to be their *greatest common divisor* and is denoted by the symbol $(a, b, \ldots, l)$.

In view of the finiteness of the number of common divisors the existence of the greatest common divisor is evident. If $(a, b, \ldots, l) = 1$, then $a, b, \ldots, l$ are said to be *relatively prime*. If each of the numbers $a, b, \ldots, l$ is relatively prime to any other of them, then $a, b, \ldots, l$ are said to be *pairwise prime*. It is evident that pairwise prime numbers are also relatively prime; in the case of two numbers the concepts of "pairwise prime" and "relatively prime" coincide.

Examples. The numbers 6, 10, 15 are relatively prime since $(6, 10, 15) = 1$. The numbers 8, 13, 21 are pairwise prime since $(8, 13) = (8, 21) = (13, 21) = 1$.

b. We first consider the common divisors of two numbers.

1. *If a is a multiple of b, then the set of common divisors of the numbers a and b coincides with the set of divisors of b; in particular,* $(a, b) = b$.

Indeed, every common divisor of the numbers a and b is a divisor of b. Conversely, if a is a multiple of b, then (**1, b, §1**) every divisor of the number b is also a divisor of the number a, i.e. it is a common divisor of the numbers a and b. Thus the set of common divisors of the numbers a and b coincides with the set of divisors of b, but since the greatest divisor of the number b is b itself, we have $(a, b) = b$.

2. *If*

$$a = bq + c,$$

then the set of common divisors of the numbers a and b coincides with the set of common divisors of the numbers b and c; in particular, $(a, b) = (b, c)$.

Indeed, the above equation shows that every common divisor of the numbers a and b divides c (**2, b, §1**) and therefore is a common divisor of the numbers b and c. Conversely, the same equation shows that every common divisor of the numbers b and c divides a and consequently is a common divisor of the numbers a and b. Therefore the common divisors of the numbers a and b are just those numbers which are also common divisors of the numbers b and c; in particular, the greatest of these divisors must also coincide, i.e. $(a, b) = (b, c)$.

3

c. In order to obtain the least common divisor as well as to deduce its most important properties, *Euclid's algorithm* is applied. The latter consists of the following process. Let a and b be positive integers. By **c**, §1, we find the sequence of equations:

$$(1) \quad \begin{cases} a = bq_2 + r_2, & 0 < r_2 < b, \\ b = r_2q_3 + r_3, & 0 < r_3 < r_2, \\ r_2 = r_3q_4 + r_4, & 0 < r_4 < r_3, \\ \hdotsfor{2} \\ r_{n-2} = r_{n-1}q_n + r_n, & 0 < r_n < r_{n-1} \\ r_{n-1} = r_n q_{n+1}, \end{cases}$$

which terminates when we obtain some $r_{n+1} = 0$. The latter must occur since the sequence $b, r_2, r_3, \ldots$ as a decreasing sequence of integers cannot contain more than b positive integers.

d. Considering the equations of (1), proceeding from the top down, **(b)** shows that the common divisors of the numbers a and b are identical with the common divisors of the numbers b and r_2, are moreover identical with the common divisors of the numbers r_2 and r_3, of the numbers r_3 and r_4, ..., of the numbers r_{n-1} and r_n, and finally with the divisors of the number r_n. Along with this, we have

$$(a, b) = (b, r_2) = (r_2, r_3) = \ldots = (r_{n-1}, r_n) = r_n.$$

We arrive at the following results.

1. *The set of common divisors of the numbers a and b coincides with the set of divisors of their greatest common divisor.*

2. *This greatest common divisor is equal to r_n*, i.e. *the last non-zero remainder in Euclid's algorithm.*

Example. We apply Euclid's algorithm to the evaluation of (525, 231). We find (the auxiliary calculations are given on the left)

4

$$\begin{array}{r|r}
525 & 231 \\
462 & 2 \\ \hline
231 & 63 \\
189 & 3 \\ \hline
63 & 42 \\
42 & 1 \\ \hline
42 & 21 \\
42 & 2 \\ \hline
\end{array}
\qquad
\begin{aligned}
525 &= 231\cdot 2 + 63, \\
231 &= 63\cdot 3 + 42, \\
63 &= 42\cdot 1 + 21, \\
42 &= 21\cdot 2.
\end{aligned}$$

Here the last positive remainder is $r_4 = 21$. This means that $(525, 231) = 21$.

e.1. *If m denotes any positive integer, we have (am, bm) $= (a, b)m$.*

2. *If δ is any common divisor of the numbers a and b, then* $\left(\dfrac{a}{\delta}, \dfrac{b}{\delta}\right) = \dfrac{(a, b)}{\delta}$; *in particular,* $\left(\dfrac{a}{(a, b)}, \dfrac{b}{(a, b)}\right) = 1$, *i.e. the quotients resulting from the division of two numbers by their greatest common divisor are relatively prime numbers.*

Indeed, multiply each of the terms of the equations (1) by m. We obtain new equations, where a, b, r_2, ..., r_n are replaced by am, bm, r_2m, ..., r_nm. Therefore $(am, bm) = r_nm$, showing that proposition 1 is true.

Applying proposition 1, we find that

$$(a, b) = \left(\frac{a}{\delta}\delta, \frac{b}{\delta}\delta\right) = \left(\frac{a}{\delta}, \frac{b}{\delta}\right)\delta;$$

and this proves proposition 2.

f.1. *If $(a, b) = 1$, then $(ac, b) = (c, b)$.*

Indeed, (ac, b) divides ac and bc, which implies (**1, d**) that it also divides (ac, bc) which is equal to c by **1, e**; but (ac, b) also divides b and therefore also divides (c, b). Conversely, (c, b) divides ac and b, and therefore also divides (ac, b). Thus (ac, b) and (c, b) divide each other and are therefore equal to one another.

2. *If $(a, b) = 1$ and ac is divisible by b, then c is divisible by b.*

5

Indeed, since $(a, b) = 1$, we have $(ac, b) = (c, b)$. But if ac is a multiple of b, then $(1, \mathbf{b})$ we have $(ac, b) = b$, which means that $(c, b) = b$, i.e. c is a multiple of b.

3. *If each* $a_1, a_2, \ldots, a_m$ *is relatively prime to each* $b_1, b_2, \ldots, b_n,$ *then the product* $a_1 a_2 \ldots a_m$ *is relatively prime to the product* $b_1 b_2 \ldots b_n.$

Indeed (theorem **1**), we have

$$(a_1 a_2 a_3 \ldots a_m, b_k) = (a_2 a_3 \ldots a_m, b_k)$$
$$= (a_3 \ldots a_m, b_k) = \ldots = (a_m, b_k) = 1,$$

and moreover, setting $a_1 a_2 \ldots a_m = A$, in the same way we find

$$(b_1 b_2 b_3 \ldots b_n, A) = (b_2 b_3 \ldots b_n, A)$$
$$= (b_3 \ldots b_n, A) = \ldots = (b_n, A) = 1.$$

g. The problem of finding the greatest common divisor of more than two numbers reduces to the same problem for two numbers. Indeed, in order to find the greatest common divisor of the numbers $a_1, a_2, \ldots, a_n$ we form the sequence of numbers:

$$(a_1, a_2) = d_2, (d_2, a_3) = d_3, (d_3, a_4) = d_4, \ldots, (d_{n-1}, a_n) = d_n.$$

The number d_n is also the greatest common divisor of all the given numbers.

Indeed $(1, \mathbf{d})$, the common divisors of the numbers a_1 and a_2 coincide with the divisors of d_2; therefore the common divisors of the numbers a_1, a_2 and a_3 coincide with the common divisors of the numbers d_2 and a_3, i.e. coincide with the divisors of d_3. Moreover, we can verify that the common divisors of the numbers a_1, a_2, a_3, a_4 coincide with the divisors of d_4, and so forth, and finally, that the common divisors of the numbers $a_1, a_2, \ldots, a_n$ coincide with the divisors of d_n. But since the largest divisor of d_n is d_n itself, it is the greatest common divisor of the numbers $a_1, a_2, \ldots, a_n.$

Considering the above proof, we can see that theorem **1, d** is true for more than two numbers also. Theorems **1, e** and **2, e** are also true, because multiplication by m or division by

δ of all the numbers $a_1, a_2, \ldots, a_n$ causes all the numbers $d_1, d_2, \ldots, d_n$ to be multiplied by m or to be divided by δ.

§3. *The Least Common Multiple*

a. Any integer which is a multiple of each of a set of given numbers, is said to be their *common multiple*. The smallest positive common multiple is called the *least common multiple*.

b. We first consider the least common multiple of two numbers. Let M be any common multiple of the integers a and b. Since it is a multiple of a, $M = ak$, where k is an integer. But M is also a multiple of b, and hence

$$\frac{ak}{b}$$

must also be an integer which, setting $(a, b) = d$, $a = a_1 d$, $b = b_1 d$, can be represented in the form $\dfrac{a_1 k}{b_1}$, where $(a_1, b_1) = 1$ (**2, e,** §2). Therefore (**2, f,** §2) k must be divisible by b_1, $k = b_1 t = \dfrac{b}{d} t$, where t is an integer. Hence

$$M = \frac{ab}{d} t.$$

Conversely, it is evident that every M of this form is a multiple of a as well as b, and therefore, this form gives all the common multiples of the numbers a and b.

The smallest positive one of these multiples, i.e. the least common multiple, is obtained for $t = 1$. It is

$$m = \frac{ab}{d}.$$

Introducing m, we can rewrite the formula we have obtained for M as:

$$M = mt.$$

The last and the next to the last equations lead to the theorems:

1. *The common multiples of two numbers are identical with the multiples of their least common multiple.*

2. *The least common multiple of two numbers is equal to their product divided by their greatest common divisor.*

c. Assume that we are now required to find the least common multiple of more than two numbers $a_1, a_2, \ldots, a_n$. Letting the symbol $m(a, b)$ denote the least common multiple of the numbers a and b, we form the sequence of numbers:

$$m(a_1, a_2) = m_2, \; m(m_2, a_3) = m_3, \; \ldots, \; m(m_{n-1}, a_n) = m_n.$$

The m_n obtained in this way will be the least common multiple of all the given numbers.

Indeed (**1, b**), the common multiples of the numbers a_1 and a_2 coincide with the multiples of m_2, and hence the common multiples of the numbers a_1, a_2 and a_3 coincide with the common multiples of m_2 and a_3, i.e. they coincide with the multiples of m_3. It is then clear that the common multiples of the numbers a_1, a_2, a_3, a_4 coincide with the multiples of m_4, and so forth, and finally, that the common multiples of the numbers a_1, a_2, $\ldots$, a_n coincide with the multiples of m_n, and since the smallest positive multiple of m_n is m_n itself, it is also the least common multiple of the numbers a_1, a_2, $\ldots$, a_n.

Considering the proof given above, we see that theorem **1, b** is also true for more than two numbers. Moreover, we have shown the validity of the following theorem:

The least common multiple of pairwise prime numbers is equal to their product.

§4. *The Relation of Euclid's Algorithm to Continued Fractions*

a. Let α be an arbitrary real number. Let q be the largest integer which does not exceed α.

For a non-integer α, we have

$$\alpha = q_1 + \frac{1}{\alpha_2}; \quad \alpha_2 > 1,$$

8

Similarly, for non-integers $\alpha_2, \ldots, \alpha_{s-1}$ we have

$$\alpha_2 = q_2 + \frac{1}{\alpha_3}; \quad \alpha_3 > 1;$$

$$\ldots \ldots \ldots \ldots \ldots \ldots \ldots \ldots \ldots \ldots$$

$$\alpha_{s-1} = q_{s-1} + \frac{1}{\alpha_s}; \quad \alpha_s > 1,$$

from which we obtain the following *development of α in a continued fraction*:

$$(1) \qquad \alpha = q_1 + \cfrac{1}{q_2 + \cfrac{1}{q_3 + \cdot}}$$

$$\cdot \quad + \cfrac{1}{q_{s-1} + \cfrac{1}{\alpha_s}} \cdot$$

If α is irrational, then it is evident that there can be no integers in the sequence $\alpha, \alpha_2, \ldots$, and the above process can be continued indefinitely.

If α is rational, then, as we shall see later (b), there will eventually be an integer in the sequence $\alpha, \alpha_2, \ldots$, and the above process will be terminated.

b. If α is an irreducible rational fraction, then the development of α in a continued fraction is closely connected with Euclid's algorithm. Indeed, we have

$$a = bq_1 + r_2; \qquad \frac{a}{b} = q_1 + \frac{r_2}{b},$$

$$b = r_2 q_2' + r_3; \qquad \frac{b}{r_2} = q_2 + \frac{r_3}{r_2},$$

$$r_2 = r_3 q_3 + r_4; \qquad \frac{r_2}{r_3} = q_3 + \frac{r_4}{r_3},$$

$$\ldots \ldots \ldots \ldots \ldots \ldots \ldots \ldots \ldots \ldots \ldots \ldots \ldots$$

9

$$r_{n-2} = r_{n-1}q_{n-1} + r_n; \quad \frac{r_{n-2}}{r_{n-1}} = q_{n-1} + \frac{r_n}{r_{n-1}},$$

$$r_{n-1} = r_n q_n; \quad \frac{r_{n-1}}{r_n} = q_n,$$

from which we find

$$\frac{a}{b} = q_1 + \cfrac{1}{q_2 + \cfrac{1}{q_3 + \cfrac{\cdot}{} \cdot \cfrac{\cdot}{+ \cfrac{1}{q_n}}}}.$$

c. The numbers $q_1, q_2, \ldots,$ which occur in the expansion of the number α in a continued fraction, are called the *partial quotients* (for the case of rational α these are, by **b**, the partial quotients of the successive divisions of the Euclidean algorithm), and the fractions

$$\delta_1 = q_1, \quad \delta_2 = q_1 + \frac{1}{q_2}, \quad \delta_3 = q_1 + \cfrac{1}{q_2 + \cfrac{1}{q_3}}, \quad \ldots$$

are called the *convergents*.

d. The very simple rule for the formation of the convergents is easily obtained by noting that $\delta_s(s > 1)$ is obtained from δ_{s-1} by replacing q_{s-1} in the expression for δ_{s-1} by $q_{s-1} + \dfrac{1}{q_s}$.

Indeed, setting $P_0 = 1$, $Q_0 = 0$, for the sake of uniformity, we can represent the convergents recursively in the following

10

way (when the equation $\dfrac{A}{B} = \dfrac{P_s}{Q_s}$ is written here, it means that A is denoted by the symbol P_s, and B by the symbol Q_s):

$$\delta_1 = \frac{q_1}{1} = \frac{P_1}{Q_1}, \quad \delta_2 = \frac{q_1 + \dfrac{1}{q_2}}{1} = \frac{q_2 q_1 + 1}{q_2 \cdot 1 + 0} = \frac{q_2 P_1 + P_0}{q_2 Q_1 + Q_0} = \frac{P_2}{Q_2}$$

$$\delta_3 = \frac{\left(q_2 + \dfrac{1}{q_3}\right) P_1 + P_0}{\left(q_2 + \dfrac{1}{q_3}\right) Q_1 + Q_0} = \frac{q_3 P_2 + P_1}{q_3 Q_2 + Q_1} = \frac{P_3}{Q_3}$$

etc., and in general

$$\delta_s = \frac{q_s P_{s-1} + P_{s-2}}{q_s Q_{s-1} + Q_{s-2}} = \frac{P_s}{Q_s}$$

Thus the numerators and the denominators of the convergents can be recursively calculated by means of the formulae

(2)
$$\begin{cases} P_s = q_s P_{s-1} + P_{s-2}, \\ Q_s = q_s Q_{s-1} + Q_{s-2}. \end{cases}$$

These calculations can easily be carried out by means of the following schema:

q_s		q_1	q_2	....			q_s	....		q_n
P_s	1	q_1	P_2	....	P_{s-2}	P_{s-1}	P_s	....	P_{n-1}	a
Q_s	0	1	Q_2	....	Q_{s-2}	Q_{s-1}	Q_s	....	Q_{n-1}	b

11

Example. Develop the number $\dfrac{105}{38}$ in a continuous fraction.

Here

```
          105 | 38
           76 | 2
       38 | 29
       29 | 1
   29 | 9
   27 | 3
 9 | 2
 8 | 4
2 | 1
2 | 2
```

$$\frac{105}{38} = 2 + \cfrac{1}{1 + \cfrac{1}{3 + \cfrac{1}{4 + \cfrac{1}{2}}}}$$

Therefore the aforementioned schema gives:

q_s		2	1	3	4	2
P_s	1	2	3	11	47	105
Q_s	0	1	1	4	17	38

e. We now consider the difference $\delta_s - \delta_{s-1}$ of successive convergents. For $s > 1$, we find

$$\delta_s - \delta_{s-1} = \frac{P_s}{Q_s} - \frac{P_{s-1}}{Q_{s-1}} = \frac{h_s}{Q_s Q_{s-1}}$$

where $h_s = P_s Q_{s-1} - Q_s P_{s-1}$; replacing P_s and Q_s by their expressions in (2) and making the evident simplifications, we find that $h_s = -h_{s-1}$. The latter, in conjunction with $h_1 = q_1 .0 - 1 \cdot 1 = -1$, gives $h_s = (-1)^s$. Thus

(3) $\qquad P_s Q_{s-1} - Q_s P_{s-1} = (-1)^s \qquad (s > 0),$

12

$$(4) \qquad \delta_s - \delta_{s-1} = \frac{(-1)^s}{Q_s Q_{s-1}} \quad (s > 1).$$

Example. In the table of the example given in **d**, we have

$$105 \cdot 17 - 38 \cdot 47 = (-1)^5 = -1.$$

f. It follows from (3) that (P_s, Q_s) divides $(-1)^s = \pm 1$ (**2, b, §1**). Hence $(P_s, Q_s) = 1$, i.e. *the convergents* $\dfrac{P_s}{Q_s}$ *are irreducible.*

g. We now investigate the sign of the difference $\delta_s - \alpha$ for δ_s which are not equal to α (i.e. we exclude the case in which δ_s is the last convergent for rational α). It is evident that δ_s is obtained by replacing α_s by q_s in the expression (1) for α. But, as is evident from **a**, as a result of this replacement

$$
\begin{array}{ll}
\alpha_s & \text{is decreased,} \\
\alpha_{s-1} & \text{is increased,} \\
\alpha_{s-2} & \text{is decreased,} \\
\multicolumn{2}{c}{\dots\dots\dots\dots\dots\dots\dots} \\
\alpha & \left\{
\begin{array}{l}
\text{is decreased for odd } s, \\[4pt]
\text{is increased for even } s.
\end{array}
\right.
\end{array}
$$

Therefore $\delta_s - \alpha < 0$ for odd s and $\delta_s - \alpha > 0$ for even s, and consequently, the sign of $\delta_s - \alpha$ coincides with the sign of $(-1)^s$.

h. *We have*

$$\left| \alpha - \delta_{s-1} \right| \leqslant \frac{1}{Q_s Q_{s-1}}$$

Indeed, for $\delta_s = \alpha$ this assertion follows (with the equality sign) from (4). For δ_s unequal to α, it follows (with the inequality sign) from (4) and from the fact that, $\delta_s - \alpha$ and $\delta_{s-1} - \alpha$ have different signs, because of **g**.

13

§5. *Prime Numbers*

a. The number 1 has only one positive divisor, namely 1. In this respect the number 1 stands alone in the sequence of natural numbers.

Every integer, greater than 1, has no fewer than two divisors, namely 1 and itself; if these divisors exhaust all the positive divisors of an integer, then it is said to be *prime*. An integer > 1 which has positive divisors other than 1 and itself, is said to be *composite*.

b. *The smallest divisor, different from one, of an integer greater than one, is a prime number.*

Indeed, let q be the smallest divisor, different from one, of the integer $a > 1$. If q were composite, then it would have some divisor q_1 such that $1 < q_1 < q$; but the number a, being divisible by q, would also be divisible by q_1 **(1, b, §1)**, and this contradicts our hypothesis concerning the number q.

c. *The smallest divisor, different from 1, of a composite number a (by* **b**, *it will be prime) does not exceed* $\sqrt{a}$.

Indeed, let q be this divisor; then $a = qa_1$, $a_1 \geqslant q$ from which, multiplying by q, we obtain $a \geqslant q^2$, $q \leqslant \sqrt{a}$.

d. *The number of primes is infinite.*

The validity of this theorem follows from the fact that no matter what different primes $p_1, p_2, \ldots, p_k$ are considered, we can obtain a new prime which is not among them. Such a prime is any prime divisor of the sum $p_1 p_2 \ldots p_k + 1$ which, dividing the whole sum, cannot be equal to any of the primes $p_1, p_2, \ldots, p_k$ **(2, b, §1)**.

e. There is a simple method, called the *sieve of Eratosthenes*, for the formation of a table of the primes not exceeding a given N. It consists of the following.

We write down the numbers

(1) $1, 2, \ldots, N.$

The first number of this sequence greater than one is 2; it is only divisible by 1 and itself, and hence it is a prime.

14

We delete from the sequence (1) (since they are composite numbers) all the numbers which are multiples of 2, except 2 itself. The first number following 2 which is not deleted is 3; it is not divisible by 2 (otherwise it would have been deleted), and hence 3 is divisible only by 1 and itself, and hence it is also prime.

Delete from the sequence (1) all the numbers which are multiples of 3, except 3 itself. The first number following 3 which is not deleted is 5; it is not divisible by either 2 or 3 (otherwise it would have been deleted). Therefore 5 is divisible only by 1 and itself, and therefore it is also prime.

And so forth.

When this process has deleted all the numbers which are multiples of primes less than the prime p, then all the numbers remaining which are less than p^2 are primes. Indeed, every composite number a which is less than p^2 has already been deleted since it is a multiple of its smallest prime divisor which is $\leqslant \sqrt{a} < p$. This implies:

1. *In the process of deleting the multiples of the prime p, this set of deleted numbers must start with p^2.*

2. *The formation of the table of primes $\leqslant N$ is completed once we have deleted all the composite multiples of primes not exceeding $\sqrt{N}$.*

§6. *The Unicity of Prime Decomposition*

a. *Every integer a is either relatively prime to a given prime p, or is divisible by p.*

Indeed, (a, p), being a divisor of p, is either 1 or p. In the first case, a is relatively prime to p, and in the second, a is divisible by p.

b. *If the product of several factors is divisible by p, then at least one of the factors is divisible by p.*

Indeed (**a**), every factor is either divisible by p or is relatively prime to p. If all the factors were relatively prime to p, then their product (**3, f, §2**) would be relatively prime to p; therefore at least one factor is divisible by p.

15

c. *Every integer greater than one can be decomposed into the product of prime factors and uniquely, if we disregard the order of the factors.*

Indeed, let a be an integer greater than unity; if p_1 is its smallest prime divisor, then $a = p_1 a_1$. If $a_1 > 1$, then if p_2 is its smallest prime divisor, we have $a_1 = p_2 a_2$. If $a_2 > 1$, then, in exactly the same way, we find $a_2 = p_3 a_3$, etc. until we come to some a_n equal to one. Then $a_{n-1} = p_n$. Multiplying all these equations together, and simplifying, we obtain the following decomposition of a into prime factors:

$$a = p_1 p_2 \cdots p_n.$$

Assume that there exists a second decomposition of the same a into prime factors $a = q_1 q_2 \cdots q_s$. Then

$$p_1 p_2 \cdots p_n = q_1 q_2 \cdots q_s.$$

The right side of this equation is divisible by q_1. Therefore **(b)**, at least one of the factors of the left side must be divisible by q_1. For example, let p_1 be divisible by q_1 (in the order of enumeration in our arrangement) then $p_1 = q_1$ (p_1 is divisible only by p_1 except for 1). Dividing both sides of the equation by $p_1 = q_1$, we have $p_2 p_3 \cdots p_n = q_2 q_3 \cdots q_s$. Repeating the preceding argumentation applied to this equation, we find $p_3 \cdots p_n = q_3 \cdots q_s$, etc., until we finally find that all the factors on one side, say the left side, are divided out. But all the factors on the right side must be cancelled simultaneously since the equation $1 = q_{n+1} \cdots q_s$ for q_{n+1}, $\ldots, q_s$ greater than 1, is impossible.

Therefore the second decomposition into prime factors is identical with the first.

d. In the decomposition of the number a into prime factors, several of them may be repeated. Letting $p_1, p_2, \ldots, p_k$ be the different primes and $\alpha_1, \alpha_2, \ldots, \alpha_k$ be the multiplicity of their occurrence in a, we obtain the so-called *canonical decomposition of a into factors*:

16

$$a = p_1^{\alpha_1} p_2^{\alpha_2} \ldots p_k^{\alpha_k}.$$

Example. The canonical decomposition of the number 588 000 is: $588\ 000 = 2^5 \cdot 3 \cdot 5^3 \cdot 7^2$.

e. Let $a = p_1^{\alpha_1} p_2^{\alpha_2} \ldots p_k^{\alpha_k}$ be the canonical decomposition of the number a. Then all the divisors of a are just all the numbers of the form

$$d = p_1^{\beta_1} p_2^{\beta_2} \ldots p_k^{\beta_k};$$

(1)

$$0 \leqslant \beta_1 \leqslant \alpha_1,\ 0 \leqslant \beta_2 \leqslant \alpha_2,\ \ldots,\ 0 \leqslant \beta_k \leqslant \alpha_k.$$

Indeed, let d divide a. Then (**b, §1**) $a = dq$, and therefore all the prime divisors of d enter into the canonical decomposition of a with indices no smaller than those with which they enter into the canonical decomposition of d. Therefore d is of the form (1).

Conversely, every d of the form (1) evidently divides a.

Example. All the divisors of the number $720 = 2^4 \cdot 3^2 \cdot 5$ can be obtained if we let β_1, β_2, β_3 in $2^{\beta_1} 3^{\beta_2} 5^{\beta_3}$ run independently through the values $\beta_1 = 0, 1, 2, 3, 4$; $\beta_2 = 0, 1, 2$; $\beta_3 = 0, 1$. Therefore these divisors are: 1, 2, 4, 8, 16, 3, 6, 12, 24, 48, 9, 18, 36, 72, 144, 5, 10, 20, 40, 80, 15, 30, 60, 120, 240, 45, 90, 180, 360, 720.

Problems for Chapter I

1. Let a and b be integers which are not both zero, and let $d = ax_0 + by_0$ be the smallest positive number of the form $ax + by$ (x and y integers). Prove that $d = (a, b)$. From this deduce theorem **1, d, §2** and the theorems of **e, §2**. Generalize these results by considering numbers of the form $ax + by + \ldots + fu$.

2. Prove that, of all the rational numbers with denominators $\leqslant Q_s$, the convergent $\delta_s = \dfrac{P_s}{Q_s}$ represents the number α most exactly.

17

3. Let the real number α be developed in a continued fraction; let N be a positive integer, let k be the number of decimal digits in it, and let n be the largest integer such that $Q_n \leqslant N$. Prove that $n \leqslant 5k + 1$. In order to prove this, compare the expressions for $Q_2, Q_3, Q_4, \ldots, Q_n$ with those which would occur if all the q_s were equal to 1, and compare the latter with the numbers $1, \xi, \xi^2, \ldots, \xi^{n-2}$ where ξ is the positive root of the equation $\xi^2 = \xi + 1$.

4. Let $r \geqslant 1$. The sequence of irreducible rational fractions with positive denominators not exceeding r, arranged in increasing order, is called the *Farey series corresponding to* r.

a. Prove that the part of the Farey series corresponding to r, containing fractions α such that $0 \leqslant \alpha \leqslant 1$, can be obtained in the following way: we write down the fractions $\dfrac{0}{1}, \dfrac{1}{1}$. If $2 \leqslant r$, then we insert the fraction $\dfrac{0 + 1}{1 + 1} = \dfrac{1}{2}$ between these fractions, and then in the resulting sequence $\dfrac{0}{1}$, $\dfrac{1}{2}, \dfrac{1}{1}$ between every two neighboring fractions $\dfrac{a_1}{b_1}$ and $\dfrac{c_1}{d_1}$ with $b_1 + d_1 \leqslant r$ we insert the fraction $\dfrac{a_1 + c_1}{b_1 + d_1}$, and so forth as long as this is possible. First prove that for any two pairs of neighboring fractions $\dfrac{a}{b}$ and $\dfrac{c}{d}$ of the sequence, obtained in the above manner, we have $ad - bc = -1$.

b. Considering the Farey series, prove the theorem: let $r \geqslant 1$, then every real number α can be represented in the form

$$\alpha = \frac{P}{Q} + \frac{\theta}{Qr}; \quad 0 < Q \leqslant r, \ (P, Q) = 1, \ |\theta| < 1.$$

18

c. Prove the theorem of problem **b** using **h**, §4.

5, a. Prove that there are an infinite number of primes of the form $4m + 3$.

b. Prove there are an infinite number of primes of the form $6m + 5$.

6. Prove that there exist an infinite number of primes by counting the number of integers, not exceeding N, whose canonical decomposition does not contain prime numbers different from $p_1, p_2, \ldots, p_k$.

7. Let K be a positive integer. Prove that the sequence of natural numbers contains an infinite set of sequences $M, M + 1, \ldots, M + K - 1$, not containing primes.

8. Prove that there are an infinite number of composite numbers among the numbers represented by the polynomial $a_0x^n + a_1x^{n-1} + \ldots + a_n$, where $n > 0$, $a_0, a_1, \ldots, a_n$ are integers and $a_0 > 0$.

9, a. Prove that the indeterminate equation (1) $x^2 + y^2 = z^2$, $x > 0, y > 0, z > 0, (x, y, z) = 1$ is satisfied by those, and only those, systems x, y, z for which one of the numbers x and y is of the form $2uv$, the other of the form $u^2 - v^2$, and finally z is of the form $u^2 + v^2$; here $u > v > 0$, $(u, v) = 1$, uv is even.

b. Using the theorem of problem **a**, prove that the equation $x^4 + y^4 = z^4$ cannot be solved in positive integers x, y, z.

10. Prove the theorem: if the equation $x^n + a_1x^{n-1} + \ldots + a_n = 0$, where $n > 0$ and $a_1, a_2, \ldots, a_n$ are integers, has a rational root then this root is an integer.

11, a. Let $S = \dfrac{1}{2} + \dfrac{1}{3} + \ldots + \dfrac{1}{n}$; $n > 1$. Prove that S is not an integer.

b. Let $S = \dfrac{1}{3} + \dfrac{1}{5} + \ldots + \dfrac{1}{2n + 1}$; $n > 0$. Prove that S is not an integer.

12. Let n be an integer, $n > 0$. Prove that all the coefficients of the expansion of the Newtonian binomial $(a + b)^n$ are odd if and only if n is of the form $2^k - 1$.

Numerical Exercises for Chapter I

1, a. Applying the Euclidean algorithm, find (6188, 4709).

b. Find (81 719, 52 003, 33 649, 30 107).

2, a. Expanding $\alpha = \dfrac{125}{92}$ in a continuous fraction and forming the table of convergents (**d, §4**), find: α) δ_4; β) the representation of α in the form considered in problem **4, b**, with $\tau = 20$.

b. Expanding $\alpha = \dfrac{5391}{3976}$ in a continuous fraction and forming the table of convergents, find: α) δ_6; β) the representation of α in the form considered in problem **4, b**, with $r = 1000$.

3. Form the Farey series (problem **4**) from 0 to 1, excluding 1, with denominators not exceeding 8.

4. Form the table of primes less than 100.

5, a. Find the canonical decomposition of the number 82 798 848.

b. Find the canonical decomposition of the number 81 057 226 635 000.

CHAPTER II

IMPORTANT NUMBER-
THEORETICAL FUNCTIONS

§1. *The Functions* [x], {x}

a. The function [x] plays an important role in number theory; it is defined for all real numbers x and is the largest integer not exceeding x. This function is called *the integral part of x.*
Examples.

$$[7] = 7; \quad [2.6] = 2; \quad [-4.75] = -5.$$

The function $\{x\} = x - [x]$ is also considered sometimes. This function is called *the fractional part of x.*
Examples.

$$\{7\} = 0; \quad \{2.6\} = 0.6; \quad \{-4.75\} = 0.25.$$

b. In order to show the usefulness of the functions we have introduced, we prove the theorem:
The power with which a given prime p enters into the product n! is equal to

$$\left[\frac{n}{p}\right] + \left[\frac{n}{p^2}\right] + \left[\frac{n}{p^3}\right] + \dots$$

Indeed, the number of factors of the product $n!$ which are

21

multiples of p is $\left[\dfrac{n}{p}\right]$; of these the number of multiples of p^2

is $\left[\dfrac{n}{p^2}\right]$; of the latter the number of multiples of p^3 is $\left[\dfrac{n}{p^3}\right]$,

etc. The sum of the latter numbers gives the required power since each factor of the product $n!$ which is a multiple of the maximal p^m is counted m times by the above process, as a multiple of p, p^2, p^3, ..., and finally, p^m.

Example. The power to which the number 3 enters into the product 40! is

$$\left[\dfrac{40}{3}\right] + \left[\dfrac{40}{9}\right] + \left[\dfrac{40}{27}\right] = 13 + 4 + 1 = 18.$$

§2. *Sums Extended over the Divisors of a Number*

a. Multiplicative functions play an important role in number theory. A function $\theta(a)$ is said to be *multiplicative* if the following conditions are satisfied:

1. *The function $\theta(a)$ is defined for all positive integers a and is not equal to zero except possibly for at most one such a.*

2. *For any two relatively prime positive integers a_1 and a_2, we have*

$$\theta(a_1 a_2) = \theta(a_1)\,\theta(a_2).$$

Example. It is not difficult to see that the function $\theta(a) = a^s$, where s is any real or complex number, is multiplicative.

b. From the aforementioned properties of the function $\theta(a)$ it follows in particular that $\theta(1) = 1$. Indeed, let $\theta(a_0)$ be different from zero, then $\theta(a_0) = \theta(1 \cdot a_0) = \theta(1)\theta(a_0)$, i.e. $\theta(1) = 1$. Moreover we have the following important property: if $\theta_1(a)$ and $\theta_2(a)$ are multiplicative functions, then $\theta_0(a) = \theta_1(a)\theta_2(a)$ is also a multiplicative function. Indeed, we find that

$$\theta_0(1) = \theta_1(1)\theta_2(1) = 1.$$

22

Moreover, for $(a_1, a_2) = 1$, we find

$$\theta_0(a_1 a_2) = \theta_1(a_1 a_2)\theta_2(a_1 a_2) = \theta_1(a_1)\theta_1(a_2)\theta_2(a_1)\theta_2(a_2) =$$
$$= \theta_1(a_1)\theta_2(a_1)\theta_1(a_2)\theta_2(a_2) = \theta_0(a_1)\theta_0(a_2).$$

c. *Let $\theta(a)$ be a multiplicative function and let $a = p_1^{\alpha_1} p_2^{\alpha_2} \ldots p_k^{\alpha_k}$ be the canonical decomposition of the number a. Then, denoting by the symbol $\displaystyle\sum_{d \backslash a}$ the sum extended over all the divisors d of the integer a, we have*

$$\sum_{d \backslash a} \theta(d) = (1 + \theta(p_1) + \theta(p_1^2) + \ldots + \theta(p_1^{\alpha_1})) \ldots$$
$$\ldots (1 + \theta(p_k) + \theta(p_k^2) + \ldots + \theta(p_k^{\alpha_k}))$$

(*if $a = 1$ the right side is considered to be equal to* 1).

In order to prove this identity, we multiply-out the right side. Then we obtain a sum of terms of the form

$$\theta(p_1^{\beta_1}) \theta(p_2^{\beta_2}) \ldots \theta(p_k^{\beta_k}) = \theta(p_1^{\beta_1} p_2^{\beta_2} \ldots p_k^{\beta_k});$$
$$0 \leqslant \beta_1 \leqslant \alpha_1, \; 0 \leqslant \beta_2 \leqslant \alpha_2, \; \ldots, \; 0 \leqslant \beta_k \leqslant \alpha_k,$$

where no terms are lacking and there are no repeated terms, and this is exactly the situation on the left (e, §6, ch. I).

d. For $\theta(a) = a^s$ the identity of **c** takes on the form

(1) $\displaystyle\sum_{d \backslash a} d^s = (1 + p_1^s + p_1^{2s} + \ldots + p_1^{\alpha_1 s}) \ldots$

$$\ldots (1 + p_k^s + p_k^{2s} + \ldots + p_k^{\alpha_k s}).$$

In particular, for $s = 1$, the left side of (1) represents *the sum of the divisors $S(a)$* of the number a. Simplifying the right side we find

$$S(a) = \frac{p_1^{\alpha_1 + 1} - 1}{p_1 - 1} \cdot \frac{p_2^{\alpha_2 + 1} - 1}{p_2 - 1} \ldots \frac{p_k^{\alpha_k + 1} - 1}{p_k - 1}.$$

23

Example.

$$S(720) = S(2^4 \cdot 3^2 \cdot 5) =$$

$$= \frac{2^{4+1} - 1}{2 - 1} \cdot \frac{3^{2+1} - 1}{3 - 1} \cdot \frac{5^{1+1} - 1}{5 - 1} = 2418.$$

For $s = 0$, the left side of (1) represents the number of divisors $\tau(a)$ of the number a and we find

$$\tau(a) = (\alpha_1 + 1)(\alpha_2 + 1) \ldots (\alpha_k + 1).$$

Example.

$$\tau(720) = (4 + 1)(2 + 1)(1 + 1) = 30.$$

§3. *The Möbius Function*

a. The *Möbius function* $\mu(a)$ is defined for all positive integers a. It is given by the equations: $\mu(a) = 0$, if a is divisible by a square different from unity; $\mu(a) = (-1)^k$ if a is not divisible by a square different from unity, where k denotes the number of prime divisors of the number a; in particular, for $a = 1$, we let $k = 0$, and hence we take $\mu(1) = 1$.

Examples.

$\mu(1) = 1,$	$\mu(5) = -1,$	$\mu(9) = 0,$
$\mu(2) = -1,$	$\mu(6) = 1,$	$\mu(10) = 1,$
$\mu(3) = -1,$	$\mu(7) = -1,$	$\mu(11) = -1,$
$\mu(4) = 0,$	$\mu(8) = 0,$	$\mu(12) = 0.$

b. *Let $\theta(a)$ be a multiplicative function and let*

$$a = p_1^{a_1} p_2^{a_2} \ldots p_k^{a_k}$$

be the canonical decomposition of the number a. Then

$$\sum_{d \backslash a} \mu(d)\theta(d) = (1 - \theta(p_1))(1 - \theta(p_2)) \ldots (1 - \theta(p_k)).$$

24

(If $a = 1$ the right side is taken to be equal to 1.)

Indeed it is evident that the function $\mu(a)$ is multiplicative. Therefore the function $\theta_1(a) = \mu(a)\theta(a)$ is also multiplicative. Applying the identity of **c, §2** to the latter, and noting that $\theta_1(p) = -\theta(p)$; $\theta_1(p^s) = 0$ for $s > 1$, we have proved the validity of our theorem.

c. In particular, setting $\theta(a) = 1$, we obtain from **b**,

$$\sum_{d \backslash a} \mu(d) \begin{cases} = 0, \text{ if } a > 1, \\ \\ = 1, \text{ if } a = 1. \end{cases}$$

Setting $\theta(d) = \dfrac{1}{d}$, we find

$$\sum_{d \backslash a} \frac{\mu(d)}{d} \begin{cases} = \left(1 - \dfrac{1}{p_1}\right)\left(1 - \dfrac{1}{p_2}\right)\cdots\left(1 - \dfrac{1}{p_k}\right), \text{ if } a > 1, \\ \\ = 1, \qquad\qquad\qquad\qquad\qquad\qquad\quad \text{ if } a = 1. \end{cases}$$

d. *Let the real or complex* $f = f_1, f_2, \ldots, f_n$ *correspond to the positive integers* $\delta = \delta_1, \delta_2, \ldots, \delta_n$. *Then, letting* S' *be the sum of the values of* f *corresponding to the values of* δ *equal to* 1, *and letting* S_d *be the sum of the values of* f *corresponding to the values of* δ *which are multiples of* d, *we have*

$$S' = \sum \mu(d)S_d,$$

where d runs through all the positive integers dividing at least one value of δ.

Indeed, in view of **c** we have

$$S' = f_1 \sum_{d \backslash \delta_1} \mu(d) + f_2 \sum_{d \backslash \delta_2} \mu(d) + \ldots + f_n \sum_{d \backslash \delta_n} \mu(d).$$

25

Gathering those terms with the same value of d and bracketing the coefficient of this $\mu(d)$, the bracket contains those and only those f whose corresponding δ are multiples of d, and this is just S_d.

§4. The Euler Function

a. *Euler's function* $\varphi(a)$ is defined for all positive integers a and represents the number of numbers of the sequence

$$(1) \qquad 0, 1, \ldots, a - 1$$

which are relatively prime to a.

Examples.

$$\begin{aligned}
\varphi(1) &= 1, & \varphi(4) &= 2, \\
\varphi(2) &= 1, & \varphi(5) &= 4, \\
\varphi(3) &= 2, & \varphi(6) &= 2.
\end{aligned}$$

b. *Let*

$$(2) \qquad a = p_1^{\alpha_1} p_2^{\alpha_2} \ldots p_k^{\alpha_k}$$

be the canonical decomposition of the number a. Then

$$(3) \qquad \varphi(a) = a \left(1 - \frac{1}{p_1}\right)\left(1 - \frac{1}{p_2}\right) \ldots \left(1 - \frac{1}{p_k}\right)$$

or also

$$(4) \qquad \varphi(a) = (p_1^{\alpha_1} - p_1^{\alpha_1 - 1})(p_2^{\alpha_2} - p_2^{\alpha_2 - 1}) \ldots (p_k^{\alpha_k} - p_k^{\alpha_k - 1});$$

in particular,

$$(5) \qquad \varphi(p^\alpha) = p^\alpha - p^{\alpha - 1}, \quad \varphi(p) = p - 1.$$

Indeed we apply the theorem of **d**, §3. Here the numbers δ and the numbers f are defined as follows: let x run through

26

the numbers of the sequence (1); to each value of x let the number $\delta = (x, a)$ and the number $f = 1$ correspond.

Then S' becomes the number of values $\delta = (x, a)$ equal to 1, i.e. becomes $\varphi(a)$. Moreover S_d becomes the number of values $\delta = (x, a)$ which are multiples of d. But (x, a) can be a multiple of d only if d is a divisor of the number a. On the strength of these conditions S_d reduces to the number of values of x which are multiples of d, i.e. to $\dfrac{a}{d}$. Thus we find

$$\varphi(a) = \sum_{d\backslash a} \mu(d)\frac{a}{d}$$

from which formula (3) follows in view of c, §3, and formula (4) follows from (3) in view of (2).

Examples.

$$\varphi(60) = 60\left(1 - \frac{1}{2}\right)\left(1 - \frac{1}{3}\right)\left(1 - \frac{1}{5}\right) = 16;$$

$$\varphi(81) = 81 - 27 = 54;$$

$$\varphi(5) = 5 - 1 = 4.$$

c. *The function* $\varphi(a)$ *is multiplicative function.*
Indeed, for $(a_1, a_2) = 1$, it follows evidently from **b** that

$$\varphi(a_1 a_2) = \varphi(a_1)\varphi(a_2).$$

Example. $\varphi(405) = \varphi(81)\varphi(5) = 54 \cdot 4 = 216.$

d. $\sum_{d\backslash a} \varphi(d) = a.$

In order to prove the validity of this formula we apply the identity of **c**, §2, which for $\theta(a) = \varphi(a)$ gives

$$\varphi(d) = (1 + \varphi(p_1) + \varphi(p_1^2) + \ldots + \varphi(p_1^{a_1}))\ldots$$
$$\ldots(1 + \varphi(p_k) + \varphi(p_k^2) + \ldots + \varphi(p_k^{a_k})),$$

27

In view of (5), the right side can be rewritten as

$$(1 + (p_1 - 1) + (p_1^2 - p_1) + \ldots + (p_1^{\alpha_1} - p_1^{\alpha_1 - 1})) \ldots$$
$$\ldots (1 + (p_k - 1) + (p_k^2 - p_k) + \ldots + (p_k^{\alpha_k} - p_k^{\alpha_k - 1})),$$

which turns out to be equal to $p_1^{\alpha_1} p_2^{\alpha_2} \ldots p_k^{\alpha_k} = a$ after gathering similar terms in each large parenthesis.

Example. Setting $a = 12$, we find

$$\varphi(1) + \varphi(2) + \varphi(3) + \varphi(4) + \varphi(6) + \varphi(12) =$$
$$= 1 + 1 + 2 + 2 + 2 + 4 = 12.$$

Problems for Chapter II

1, a. Let the function $f(x)$ be continuous and non-negative in the interval $Q \leqslant x \leqslant R$. Prove that the sum

$$\sum_{Q < x \leqslant R} [f(x)]$$

is equal to the number of lattice points (points with integer coordinates) in the plane region: $Q < x \leqslant R$, $0 < y \leqslant f(x)$.

b. Let P and Q be positive odd relatively prime integers. Prove that

$$\sum_{0 < x < \frac{P}{2}} \left[\frac{P}{Q} x \right] + \sum_{0 < y < \frac{Q}{2}} \left[\frac{Q}{P} y \right] = \frac{P - 1}{2} \cdot \frac{Q - 1}{2}.$$

c. Let $r > 0$ and let T be the number of lattice points in the region $x^2 + y^2 \leqslant r^2$. Prove that

$$T = 1 + 4[r] + 8 \sum_{0 < x \leqslant \frac{r}{\sqrt{2}}} [\sqrt{r^2 - x^2}] - 4 \left[\frac{r}{\sqrt{2}} \right]^2.$$

d. Let $n > 0$ and let T be the number of lattice points of the region $x > 0$, $y > 0$, $xy \leqslant n$. Prove that

$$T = 2 \sum_{0 < x \leqslant \sqrt{n}} \left[\frac{n}{x} \right] - [\sqrt{n}]^2 .$$

2. Let $n > 0$, m an integer, $m > 1$, and let x run through the positive integers which are not divisible by the m-th power of an integer exceeding 1. Prove that

$$\sum_x \left[\sqrt[m]{\frac{n}{x}} \right] = [n].$$

3. Let the positive numbers α and β be such that

$$[\alpha x], \quad x = 1, 2, \ldots; \quad [\beta y], \quad y = 1, 2, \ldots$$

form, taken together, all the natural numbers without repetitions. Prove that this occurs if and only if α is irrational and

$$\frac{1}{\alpha} + \frac{1}{\beta} = 1.$$

4, a. Let $r \geqslant 1$, $t = [r]$, and let $x_1, x_2, \ldots, x_t$ be the numbers $1, 2, \ldots, t$ in some order so that the numbers

$$0, \ \{\alpha x_1\}, \ \{\alpha x_2\}, \ \ldots, \ \{\alpha x_t\}, \ 1$$

are non-decreasing. Prove the theorem of problem **4, b, ch. I,** by considering the differences of neighboring numbers of the latter sequence.

b. Let $X, Y, \ldots, Z$ be real numbers, each of which is not less than 1; let $\alpha, \beta, \ldots, \gamma$ be real numbers. Prove that there exist integers $x, y, \ldots, z$, not all zero, and an integer u, satisfying the conditions:

$$|x| \leqslant X, \quad |y| \leqslant Y, \quad \ldots, \quad |z| \leqslant Z,$$

$$(x, y, \ldots, z) = 1, \quad |\alpha x + \beta y + \ldots + \gamma z - u| < \frac{1}{XY \ldots Z}$$

5. Let α be a real number, c an integer, $c > 0$. Prove that

$$\left[\frac{[\alpha]}{c}\right] = \left[\frac{\alpha}{c}\right]$$

6, a. Let $\alpha, \beta, \ldots, \lambda$ be real numbers. Prove that

$$[\alpha + \beta + \ldots + \lambda] \geqslant [\alpha] + [\beta] + \ldots + [\lambda].$$

b. Let $a, b, \ldots, l$ be positive integers, and let $a + b + \ldots$
$\ldots + l = n$. Applying **b**, §1, prove that

$$\frac{n!}{a!b! \ldots l!}$$

is an integer.

7. Let h be a positive integer, p a prime and

$$u_s = \frac{p^{s+1} - 1}{p - 1}$$

Representing h in the form $h = p_m u_m + p_{m-1} u_{m-1} + \ldots$
$\ldots + p_1 u_1 + p_0$, where u_m is the largest u_s not exceeding h,
$p_m u_m$ is the largest multiple of u_m which does not exceed h,
$p_{m-1} u_{m-1}$ is the largest multiple of u_{m-1} which does not exceed $h - p_m u_m$, $p_{m-2} u_{m-2}$ is the largest multiple of u_{m-2}
which does not exceed $h - p_m u_m - p_{m-1} u_{m-1}$ etc., prove that
numbers a such that the number p enters into the canonical
representation of $a!$ with the power h, exist if and only if all
the $p_m, p_{m-1}, \ldots, p_1, p_0$ are less than p, while, if this occurs,
the numbers a are just all the numbers of the form

$$a = p_m p^{m+1} + p_{m-1} p^m + \ldots + p_1 p^2 + p_0 p + p',$$

where p' has the values $0, 1, \ldots, p - 1$.

8, a. Let the function $f(x)$ have a continuous second derivative in the interval $Q \leqslant x \leqslant R$. Setting

$$\rho(x) = \frac{1}{2} - \{x\}, \quad \sigma(x) = \int_0^x \rho(z)dz,$$

prove (Sonin's formula)

$$\sum_{Q < x \leqslant R} f(x) = \int_Q^R f(x)dx + \rho(R)f(R) - \rho(Q)f(Q) -$$

$$- \sigma(R)f'(R) + \sigma(Q)f'(Q) + \int_Q^R \sigma(x)f''(x)dx.$$

b. Let the conditions of problem a be satisfied for arbitrarily large R, while $\displaystyle\int_R^\infty |f''(x)|\, dx$ converges. Prove that

$$\sum_{Q < x \leqslant R} f(x) = C + \int_Q^R f(x)dx + \rho(R)f(R) -$$

$$- \sigma(R)f'(R) - \int_R^\infty \sigma(x)f''(x)dx,$$

where C does not depend on R.

c. If B takes on only positive values and the ratio $\dfrac{|A|}{B}$ is bounded above, then we write $A = O(B)$.

Let n be an integer, $n > 1$. Prove that

$$\ln(n!) = n \ln n - n + O(\ln n)$$

31

9, a. Let $n \geqslant 2$, $\Theta(z, z_0) = \sum\limits_{z_0 < p \leqslant z} \ln p$, where p runs through the primes. Moreover, let $\Theta(z) = \Theta(z, 0)$ and for $x > 0$,

$$\psi(x) = \Theta(x) + \Theta(\sqrt{x}) + \Theta(\sqrt[3]{x}) + \ldots$$

Prove that

$\alpha)$ $\ln([n]!) = \psi(n) + \psi\left(\dfrac{n}{2}\right) + \psi\left(\dfrac{n}{3}\right) + \ldots$

$\beta)$ $\psi(n) < 2n$

$\gamma)$ $\Theta\left(n, \dfrac{n}{2}\right) + \Theta\left(\dfrac{n}{3}, \dfrac{n}{4}\right) + \Theta\left(\dfrac{n}{5}, \dfrac{n}{6}\right) + \ldots =$

$$= n \ln 2 + O(\sqrt{n}).$$

b. For $n > 2$, prove that

$$\sum_{p \leqslant n} \frac{\ln p}{p} = \ln n + O(1),$$

where p runs through the primes.

c. Let ϵ be an arbitrary positive constant. Prove that the sequence of natural numbers contains an infinite number of pairs p_n, p_{n+1} of prime numbers such that

$$p_{n+1} < p_n(1 + \epsilon).$$

d. Let $n > 2$. Prove that

$$\sum_{p \leqslant n} \frac{1}{p} = C + \ln \ln n + O\left(\frac{1}{\ln n}\right),$$

where p runs through the primes and C does not depend on n.

32

e. Let $n > 2$. Prove that

$$\prod_{p \leqslant n} \left(1 - \frac{1}{p}\right) = \frac{C_0}{\ln n}\left(1 + O\left(\frac{1}{\ln n}\right)\right)$$

where p runs through the primes and C_0 does not depend on n.

10, a. Let $\theta(a)$ be a multiplicative function. Prove that $\theta_1(a) = \sum_{d \backslash a} \theta(d)$ is also a multiplicative function.

b. Let the function $\theta(a)$ be defined for all positive integers a and let the function $\psi(a) = \sum_{d \backslash a} \theta(a)$ be multiplicative.

Prove that the function $\theta(a)$ is also multiplicative.

11. For $m > 0$, let $\tau_m(a)$ denote the number of solutions of the indeterminate equation $x_1 x_2 \ldots x_m = a$ ($x_1, x_2, \ldots, x_m$ run through the positive integers independently of one-another); in particular, it is evident that $\tau_1(a) = 1$, $\tau_2(a) = \tau(a)$. Prove that

a. $\tau_m(a)$ is a multiplicative function.

b. If the canonical decomposition of the number a is of the form $a = p_1 p_2 p_3 \ldots p_k$, then $\tau_m(a) = m^k$.

c. If ϵ is an arbitrary positive constant, then

$$\lim_{a \to \infty} \frac{\tau_m(a)}{a^\epsilon} = 0.$$

d. $\sum_{0 < a \leqslant n} \tau_m(a)$ is equal to the number of solutions of the inequality $x_1 x_2 \ldots x_m \leqslant a$ in positive integers $x_1, x_2, \ldots, x_m$.

12. Let $R(s)$ be the real part of the complex number s. For $R(s) > 1$, we set $\zeta(s) = \sum_{n=1}^{\infty} \frac{1}{n^s}$. Let m be a positive integer.

Prove that

$$(\zeta(s))^m = \sum_{n=1}^{\infty} \frac{\tau_m(n)}{n^s}$$

13, a. For $R(s) > 1$, prove that

$$\zeta(s) = \prod \frac{1}{1 - \dfrac{1}{p^s}}$$

where p runs through all the primes.

b. Prove that there exist an infinite number of primes, starting from the fact that the harmonic series diverges.

c. Prove that there exist an infinite number of primes, starting from the fact that $\zeta(2) = \dfrac{\pi^2}{6}$ is an irrational number.

14. Let $\Lambda(a) = \ln p$ for $a = p^l$, where p is a prime and l is a positive integer; and let $\Lambda(a) = 0$ for all other positive integers a. For $R(s) > 1$, prove that

$$\frac{\zeta'(s)}{\zeta(s)} = -\sum_{n=1}^{\infty} \frac{\Lambda(n)}{n^s}$$

15. Let $R(s) > 1$. Prove that

$$\prod_{p} \left(1 - \frac{1}{p^s}\right) = \sum_{n=1}^{\infty} \frac{\mu(n)}{n^s}$$

where p runs through all the primes.

16, a. Let $n \geqslant 1$. Applying **d, §3**, prove that

$$1 = \sum_{0 < d \leqslant n} \mu(d) \left[\frac{n}{d}\right].$$

b. Let $M(z, z_0) = \sum_{z_0 < a \leqslant z} \mu(a)$; $M(x) = M(x, 0)$. Prove that

$\alpha)$ $M(n) + M\left(\dfrac{n}{2}\right) + M\left(\dfrac{n}{3}\right) + \ldots = 1,\ n \geqslant 1.$

34

$\beta)$ $M\left(n, \dfrac{n}{2}\right) + M\left(\dfrac{n}{3}, \dfrac{n}{4}\right) + M\left(\dfrac{n}{5}, \dfrac{n}{6}\right) + \ldots = -1, n \geqslant 2.$

c. Let $n \geqslant 1$, let l be an integer, $l > 1$, and let $T_{l,n}$ be the number of integers x, such that $0 < x \leqslant n$, which are not divisible by the l-th power of an integer exceeding 1. Applying **d, §3**, prove that

$$T_{l,n} = \sum_{d=1}^{\infty} \mu(d)\left[\dfrac{n}{d^l}\right].$$

17, a. Let a be a positive integer and let the function $f(x)$ be uniquely defined for the integers $x_1, x_2, \ldots, x_n$. Prove

$$S' = \sum_{d\backslash a} \mu(d)S_d,$$

where S' is the sum of the values of $f(x)$ extended over those values of x which are relatively prime to a, and S_d is the sum of the values of $f(x)$ extended over those values of x which are multiples of d.

b. Let $k > 1$ and consider the systems

$$x_1', x_2', \ldots, x_k'; x_1'', x_2'', \ldots, x_k''; \ldots; x_1^{(n)}, x_2^{(n)}, \ldots, x_k^{(n)},$$

each of which consists of integers, not all zero. Moreover, let the function $f(x_1, x_2, \ldots, x_k)$ be uniquely defined for these systems. Prove that

$$S' = \sum \mu(d)S_d,$$

where S' is the sum of the values of $f(x_1, x_2, \ldots, x_k)$ extended over systems of relative prime numbers, and S_d is the sum of the values of $f(x_1, x_2, \ldots, x_k)$ extended over systems of numbers which are all multiples of d. Here d runs through positive integers.

35

c. Let a be a positive integer, and let $F(\delta)$ be uniquely defined for the divisors δ of the number a. Setting

$$G(\delta) = \sum_{d \backslash \delta} F(d),$$

prove (the inversion law for number-theoretic functions)

$$F(a) = \sum_{d \backslash a} \mu(d) G\left(\frac{a}{d}\right).$$

d. Associate with the positive integers

$$\delta_1, \ \delta_2, \ \dots, \ \delta_n$$

arbitrary real or complex numbers

$$f_1, \ f_2, \ \dots, \ f_n$$

different from zero. Prove that

$$P' = \prod P_d^{\mu(d)}$$

where P' is the product of the values f associated with values of δ equal to one, and P_d is the product of the values f associated with values of δ which are multiples of d, where d runs through all the positive integers which divide at least one δ.

18. Let a be an integer, $a > 1$, $\sigma_m(n) = 1^m + 2^m + \dots + n^m$; let $\psi_m(a)$ be the sum of the m-th powers of the numbers of the sequence $1, 2, \dots, a$ which are relatively prime to a; let $p_1, p_2, \dots, p_k$ be all the prime divisors of the number a.

a. Applying the theorem of problem **17, a,** prove that

$$\psi_m(a) = \sum_{d \backslash a} \mu(d) d^m \sigma_m\left(\frac{a}{d}\right).$$

b. Prove that

$$\psi_1(a) = \frac{a}{2}\, \varphi(a).$$

c. Prove that

$$\psi_2(a) = \left(\frac{a^2}{3} + \frac{(-1)^k}{6} p_1 p_2 \cdots p_k \right) \varphi(a).$$

19. Let $z > 1$, let a be a positive integer; let T_z be the number of numbers x such that $0 < x \leqslant z$, $(x, a) = 1$; let ϵ be an arbitrary positive constant.

a. Prove that

$$T_z = \sum_{d \backslash a} \mu(d) \left[\frac{z}{d} \right].$$

b. Prove that

$$T_z = \frac{z}{a}\, \varphi(a) + O(a^\epsilon).$$

c. Let $z > 1$; let $\pi(z)$ be the number of prime numbers not exceeding z; let a be the product of the primes not exceeding $\sqrt{z}$. Prove that

$$\pi(z) = \pi(\sqrt{z}) - 1 + \sum_{d \backslash a} \mu(d) \left[\frac{z}{d} \right].$$

20. Let $R(s) > 1$ and let a be a positive integer. Prove that

$$\sum{}' \frac{1}{n^s} = \prod \left(1 - \frac{1}{p^s} \right) \zeta(s),$$

where, on the left side, n runs through the positive integers relatively prime to a, while, on the right side, p runs through all the prime divisors of the number a.

21, a. The probability P that k positive integers $x_1, x_2, \ldots, x_k$ are relatively prime is defined as the limit, as $N \to \infty$, of the probability P_N that the k numbers $x_1, x_2, \ldots, x_k$ are relatively prime, when these k numbers take on the values 1, 2, $\ldots, N$ independently and with equal probability. Applying the theorem of problem **17, b,** prove that $P = (\zeta(k))^{-1}$.

b. Defining the probability of the irreducibility of the fraction $\dfrac{x}{y}$ as in problem **a** for $k = 2$, prove that $P = \dfrac{6}{\pi^2}$.

22, a. Let $r > 2$ and let T be the number of lattice points (x, y) with relatively prime coordinates in the region $x^2 + y^2 \leqslant r$. Prove that

$$T = \frac{6}{\pi} r^2 + O(r \ln r).$$

b. Let $r \geqslant 2$ and let T be the number of lattice points (x, y, z) with relatively prime coordinates lying in the region $x^2 + y^2 + z^2 \leqslant r^2$. Prove that

$$T = \frac{4\pi}{3\zeta(3)} r^3 + O(r^2)$$

23, a. Prove the first theorem of **c, §3,** by considering the divisors of the number a which are not divisible by the square of an integer exceeding 1, and having 1, 2, $\ldots$ prime divisors.

b. Let a be an integer, $a > 1$, and let d run through the divisors of the number a having no more than m prime divisors; Prove that $\sum \mu(d) \geqslant 0$ for m even, and $\sum \mu(d) \leqslant 0$ for m odd.

c. Under the conditions of the theorem of **d, §3,** assuming all the f to be non-negative and letting d run only through the numbers having no more than m prime divisors, prove that

$$S' \leqslant \sum \mu(d) S_d, \quad S' \geqslant \sum \mu(d) S_d$$

according as m is even or odd.

d. Prove the validity of the same inequalities as in problem **c**, under the conditions of problem **17, a**, assuming all the values of $f(x)$ are non-negative, as well as under the conditions of **17, b**, assuming all the values of $f(x_1, x_2, \ldots, x_k)$ are non-negative.

24. Let ϵ be an arbitrary constant such that $0 < \epsilon < \dfrac{1}{6}$; let N be an integer, $r = \ln N$, $0 < q \leqslant N^{1-\epsilon}$, $0 \leqslant l < q$, $(q, l) = 1$; let $\pi(N, q, l)$ be the number of primes such that $p \leqslant N$, $p = qt + l$, where t is an integer. Prove that

$$\pi(N, q, l) = O(\Delta); \quad \Delta = \frac{N(qr)^\epsilon}{qr}$$

In order to prove this, setting $h = r^{1-\epsilon}$, the primes satisfying the above condition can be considered to be among all numbers satisfying these conditions relatively prime to a, where a is the product of all primes which do not exceed e^h and do not divide q. We can then apply the theorem of problem **23, d** (under the conditions of problem **17, a**) with the above a and $m = 2[2\ln r + 1]$.

25. Let k be a positive even number, let the canonical decomposition of the number a be of the form $a = p_1 p_2 \ldots p_k$ and let d run through the divisors of the number a such that $0 < d < \sqrt{a}$. Prove that

$$\sum_d \mu(d) = 0.$$

26. Let k be a positive integer, let d run through the positive integers such that $\varphi(d) = k$. Prove that

$$\sum \mu(d) = 0.$$

27. Using the expression for $\varphi(a)$, prove that there exist an infinite number of primes.

39

28, a. Prove the theorem of **d, §4** by showing that the number of integers of the sequence 1, 2, ..., a which have the same greatest common divisor δ with a, is equal to $\varphi\left(\dfrac{a}{\delta}\right)$.

b. Deduce expressions for $\varphi(a)$:

        α) using the theorem of problem **10, b**;

        β) using the theorem of problem **17, c**.

29. Let $R(s) > 2$. Prove that

$$\sum_{n=1}^{\infty} \frac{\varphi(n)}{n^s} = \frac{\zeta(s-1)}{\zeta(s)}$$

30. Let n be an integer, $n \geqslant 2$. Prove that

$$\sum_{m=1}^{n} \varphi(m) = \frac{3}{\pi^2} n^2 + O(n \ln n).$$

Numerical Exercises for Chapter II

1, a. Find the exact power with which 5 enters into the canonical decomposition of 5258! (problem **5**).

b. Find the canonical decomposition of the number 125!

2, a. Find $\tau(2\ 800)$ and $S(2\ 800)$.

b. Find $\tau(232\ 848)$ and $S(232\ 848)$.

3. Form the table of values of the function $\mu(a)$ for all $a = 1, 2, \ldots, 100$.

4. Find α) $\varphi(5040)$; β) $\varphi(1\ 294\ 700)$.

5. Form the table of values of the function $\varphi(a)$ for all $a = 1, 2, \ldots, 50$, using only formula (5), §4, and theorem **c, §4**.

CHAPTER III

CONGRUENCES

§1. *Basic Concepts*

a. We will consider integers in relation to the remainders resulting from their division by a given positive integer m which we call the *modulus*.

To each integer corresponds a unique remainder resulting from its division by m (**c, §1, ch. I**); if the same remainder r corresponds to two integers a and b, then they are said to be *congruent* modulo m.

b. The congruence of the numbers a and b modulo m is written as

$$a \equiv b(\text{mod } m),$$

which is read: a is congruent to b modulo m.

c. *The congruence of the numbers a and b modulo m is equivalent to:*

1. *The possibility of representing a in the form* $a = b + mt$, *where t is an integer.*

2. *The divisibility of* $a - b$ *by m.*

Indeed, it follows from $a \equiv b(\text{mod } m)$ that

$$a = mq + r, \ b = mq_1 + r; \ 0 \leqslant r < m,$$

and hence

$$a - b = m(q - q_1), \ a = b + mt, \ t = (q - q_1).$$

41

Conversely, from $a = b + mt$, representing b in the form

$$b = mq_1 + r, \ 0 \leqslant r < m,$$

we deduce

$$a = mq + r; \ q = q_1 + t,$$

i.e.

$$a \equiv b(\text{mod } m)$$

proving assertion 1.

Assertion 2 follows immediately from assertion 1.

§2. *Properties of Congruences similar to those of Equations*

a. *Two numbers which are congruent to a third are congruent to each other.*

This follows from **a, §1**.

b. *Congruences can be added termwise.*

Indeed, let

(1) $a_1 \equiv b_1(\text{mod } m), \ a_2 \equiv b_2(\text{mod } m), \ \ldots, \ a_k \equiv b_k(\text{mod } m)$

Then (**1, c, §1**)

(2) $\quad a_1 = b_1 + mt_1, \ a_2 = b_2 + mt_2, \ \ldots, \ a_k = b_k + mt_k,$

and hence

$$a_1 + a_2 + \ldots + a_k = b_1 + b_2 + \ldots + b_k + m(t_1 + t_2 + \ldots + t_k),$$

or (**1, c, §1**)

$$a_1 + a_2 + \ldots + a_k \equiv b_1 + b_2 + \ldots + b_k(\text{mod } m)$$

A summand on either side of a congruence can be put on the other side by changing its sign.

Indeed, adding the congruence $a + b \equiv c \pmod{m}$ to the evident congruence $-b \equiv -b \pmod{m}$, we find $a \equiv c - b \pmod{m}$.

Any number which is a multiple of the modulus can be added to (or subtracted from) any side of a congruence.

Indeed, adding the congruence $a \equiv b \pmod{m}$ to the evident congruence $mk \equiv 0 \pmod{m}$, we obtain $a + mk \equiv b \pmod{m}$.

c. *Congruences can be multiplied termwise.*

Indeed, we again consider the congruences (1) and deduce from them the equations (2). Multiplying equations (2) together termwise we find

$$a_1 a_2 \ldots a_k = b_1 b_2 \ldots b_k + mN,$$

where N is an integer. Consequently (1, c, § 1),

$$a_1 a_2 \ldots a_k \equiv b_1 b_2 \ldots b_k \pmod{m}.$$

Both sides of a congruence can be raised to the same power.
This follows from the preceding theorem.

Both sides of a congruence can be multiplied by the same integer.

Indeed, mutliplying the congruence $a \equiv b \pmod{m}$ by the evident congruence $k \equiv k \pmod{m}$, we find $ak \equiv bk \pmod{m}$.

d. Properties b and c (addition and multiplication of congruences) can be generalized to the following theorem.

If we replace A, x_1, x_2, ..., x_k in the expression of an integral rational function $S = \sum A x_1^{a_1} x_2^{a_2} \ldots x_k^{a_k}$ with integral coefficients, by the numbers B, y_1, y_2, ..., y_k which are congruent to the preceding ones modulo m, then the new expression S will be congruent to the old one modulo m.

Indeed, from

$$A \equiv B \pmod{m}, \quad x_1 \equiv y_1 \pmod{m},$$

$$x_2 \equiv y_2 \pmod{m}, \quad \ldots, \quad x_k \equiv y_k \pmod{m}$$

we find (c)

$$A \equiv B(\text{mod } m), \quad x_1^{a_1} \equiv y_1^{a_1}(\text{mod } m)$$
$$x_2^{a_2} \equiv y_2^{a_2}(\text{mod } m), \quad \ldots, \quad x_k^{a_k} \equiv y_k^{a_k}(\text{mod } m),$$
$$A x_1^{a_1} x_2^{a_2} \ldots x_k^{a_k} \equiv B y_1^{a_1} y_2^{a_2} \ldots y_k^{a_k}(\text{mod } m)$$

from which, summing, we find

$$\sum A x_1^{a_1} x_2^{a_2} \ldots x_k^{a_k} = \sum B y_1^{a_1} y_2^{a_2} \ldots y_k^{a_k}(\text{mod } m).$$

If

$$a \equiv b(\text{mod } m), \quad a_1 \equiv b_1(\text{mod } m), \quad \ldots, \quad a_n \equiv b_n(\text{mod } m),$$
$$x \equiv x_1(\text{mod } m),$$

then

$$ax^n + a_1 x^{n-1} + \ldots + a_n \equiv b x_1^n + b_1 x_1^{n-1} + \ldots + b_n(\text{mod } m).$$

This result is a special case of the preceding one.

e. *Both sides of a congruence can be divided by one of their common divisors if it is relatively prime to the modulus.*

Indeed, it follows from $a \equiv b(\text{mod } m)$, $a = a_1 d$, $b = b_1 d$, $(d, m) = 1$ that the difference $a - b$, which is equal to $(a_1 - b_1)d$, is divisible by m. Therefore (2, **f**, **§2, ch. I**) $a_1 - b_1$ is divisible by m, i.e. $a_1 \equiv b_1(\text{mod } m)$.

§3. *Further Properties of Congruences*

a. *Both sides of a congruence and the modulus can be multiplied by the same integer.*

Indeed, it follows from $a \equiv b(\text{mod } m)$ that

$$a = b + mt, \quad ak = bk + mkt$$

and hence, $ak \equiv bk(\text{mod } mk)$.

44

b. *Both sides of a congruence and the modulus can be divided by any one of their common divisors.*

Indeed, let

$$a \equiv b(\mathrm{mod}\ m), \quad a = a_1 d, \quad b = b_1 d, \quad m = m_1 d.$$

We have

$$a = b + mt, \quad a_1 d = b_1 d + m_1 dt, \quad a_1 = b_1 + m_1 t$$

and hence $a_1 \equiv b_1(\mathrm{mod}\ m_1)$.

c. *If the congruence $a \equiv b$ holds for several moduli, then it also holds for the modulus equal to the least common multiple of these moduli.*

Indeed, it follows from $a \equiv b(\mathrm{mod}\ m_1)$, $a \equiv b(\mathrm{mod}\ m_2)$, ..., $a \equiv b(\mathrm{mod}\ m_k)$ that the difference $a - b$ is divisible by all the moduli $m_1, m_2, \ldots, m_k$. Therefore (c, §3, ch. I) it must be divisible by the least common multiple m of these moduli, i.e. $a \equiv b(\mathrm{mod}\ m)$.

d. *If a congruence holds modulo m, then it also holds modulo d, which is equal to any divisor of the number m.*

Indeed, it follows from $a \equiv b(\mathrm{mod}\ m)$ that the difference $a - b$ must be divisible by m; therefore (1, b, §1, ch. I) it must be divisible by any divisor d of the number m, i.e. $a \equiv b(\mathrm{mod}\ d)$.

e. *If one side of a congruence and the modulus are divisible by some number then the other side of the congruence must also be divisible by the same number.*

Indeed, it follows from $a \equiv b(\mathrm{mod}\ m)$ that $a = b + mt$, and if a and m are multiples of d, then (2, b, §1, ch. I) b must also be a multiple of d, as was to be proven.

f. *If $a \equiv b(\mathrm{mod}\ m)$, then $(a, m) = (b, m)$.*

Indeed, in view of 2, b, §2, ch. I this equation follows immediately from $a = b + mt$.

§4. *Complete Systems of Residues*

a. Numbers which are congruent modulo m form an *equivalence class modulo m.*

45

It follows from this definition that all the numbers of an equivalence class have the same remainder r, and we obtain all the numbers of an equivalence class if we let q in the form $mq + r$ run through all the integers.

Corresponding to the m different values of r we have m equivalence classes of numbers modulo m.

b. Any number of an equivalence class is said to be a *residue* modulo m with respect to all the numbers of the equivalence class. The residue obtained for $q = 0$ is equal to the remainder r itself, and is called the *least non-negative residue*.

The residue ρ of smallest absolute value is called *the absolutely least residue*.

It is evident that we have $\rho = r$ for $r < \dfrac{m}{2}$; for $r > \dfrac{m}{2}$ we have $\rho = r - m$; finally, if m is even and $r = \dfrac{m}{2}$, then we can take for ρ either of the two numbers $\dfrac{m}{2}$ and $\dfrac{m}{2} - m = -\dfrac{m}{2}$.

Taking one residue from each equivalence class, we obtain a *complete system of residues modulo m*. Frequently, as a complete system of residues we use the least non-negative residues $0, 1, \ldots, m - 1$ or the absolutely least residues; the latter, as follows from our above discussion, is represented in the case of odd m by the sequence

$$-\frac{m-1}{2}, \ldots, -1, 0, 1, \ldots, \frac{m-1}{2},$$

and in the case of even m by either of the two sequences

$$-\frac{m}{2} + 1, \ldots, -1, 0, 1, \ldots, \frac{m}{2},$$

$$-\frac{m}{2}, \ldots, -1, 0, 1, \ldots, \frac{m}{2} + 1.$$

c. *Any m numbers which are pairwise incongruent modulo m form a complete system of residues modulo m.*

Indeed, being incongruent, these numbers must belong to different equivalence classes, and since there are m of them, i.e. as many as there are classes, it follows that one number falls into each class.

d. *If $(a, m) = 1$ and x runs over a complete system of residues modulo m, then $ax + b$, where b is any integer also runs over a complete system of residues modulo m.*

Indeed, there are as many numbers $ax + b$ as there are numbers x, i.e. m. Accordingly, it only remains to prove that any two numbers $ax_1 + b$ and $ax_2 + b$ corresponding to incongruent x_1 and x_2 will also be incongruent modulo m.

But, assuming that $ax_1 + b \equiv ax_2 + b(\mathrm{mod}\ m)$, we arrive at the congruence $ax_1 \equiv ax_2 \pmod{m}$, from which we obtain $x_1 \equiv x_2 \pmod{m}$ as a consequence of $(a, m) = 1$, and this contradicts the assumption of the incongruence of the numbers x_1 and x_2.

§5. *Reduced Systems of Residues*

a. By **f**, §3, the numbers of an equivalence class modulo m all have the same greatest common divisor relative to the modulus. Particularly important are the equivalence classes for which this divisor is equal to unity, i.e. the classes containing numbers relatively prime to the modulus.

Taking one residue from each such class we obtain a *reduced system of residues modulo m*. A reduced system of residues therefore consists of the numbers of a complete system which are relatively prime to the modulus. A reduced system of residues is usually chosen from among the numbers of the system of least non-negative residues $0, 1, \ldots, m - 1$. Since the number of these numbers which are relatively prime to m is $\varphi(m)$, the number of numbers of a reduced system, which is equal to the number of equivalence classes containing numbers relatively prime to the modulus, is $\varphi(m)$.

Example. A reduced system of residues modulo 42 is

$$1, 5, 11, 13, 17, 19, 23, 25, 29, 31, 37, 41.$$

b. *Any* $\varphi(m)$ *numbers which are pairwise incongruent modulo m and relatively prime to the modulus form a reduced system of residues modulo m.*

Indeed, being incongruent and relatively prime to the modulus, these numbers belong to different equivalence classes which contain numbers relatively prime to the modulus, and since there are $\varphi(m)$ of them, i.e. as many as there are classes of the above kind, it follows that there is one number in each class.

c. *If* $(a, m) = 1$ *and x runs through a reduced system of residues modulo m, then ax also runs through a reduced system of residues modulo m.*

Indeed, there are as many numbers ax as there are numbers x, i.e. $\varphi(m)$. By **b**, it only remains to prove that the numbers ax are incongruent modulo m and are relatively prime to the modulus. But the first was proved in **d**, §4 for the numbers of the more general form $ax + b$, and the second follows from $(a, m) = 1$, $(x, m) = 1$.

§6. The Theorems of Euler and Fermat

a. *For* $m > 1$ *and* $(a, m) = 1$, *we have (Euler's theorem):*

$$a^{\varphi(m)} \equiv 1 \pmod{m}.$$

Indeed, if x runs through a reduced system of residues

$$x = r_1, r_2, \ldots, r_c; \quad c = \varphi(m),$$

which consists of the least non-negative residues, then the least non-negative residues $\rho_1, \rho_2, \ldots, \rho_c$ of the numbers ax will run through the same system, but, generally speaking, in a different order (**c**, §**5**).

48

Multiplying the congruences

$$ar_1 \equiv \rho_1 \pmod{m}, \ ar_2 \equiv \rho_2 \pmod{m}, \ \ldots, \ ar_c \equiv \rho_c \pmod{m}$$

together termwise, we find

$$a^c r_1 r_2 \ldots r_c \equiv \rho_1 \rho_2 \ldots \rho_c \pmod{m},$$

from which we find

$$a^c \equiv 1 \pmod{m}$$

by dividing both sides by the product $r_1 r_2 \ldots r_c = \rho_1 \rho_2 \ldots \rho_c$.

b. *If p is a prime and a is not divisible by p, then we have (Fermat's theorem):*

(1) $$a^{p-1} \equiv 1 \pmod{p}.$$

This theorem is a consequence of theorem **a** for $m = p$. The latter theorem can be put in better form. Indeed, multiplying both sides of the congruence (1) by a, we obtain the congruence

$$a^p \equiv a \pmod{p},$$

which is valid for all integers a, since it is valid for integers a which are multiples of p.

Problems for Chapter III

1, a. Representing an integer in the ordinary decimal system, deduce criteria for divisibility by 3, 9, 11.

b. Representing an integer in the calculational system to the base 100, deduce a criterion for divisibility by 101.

c. Representing an integer in the calculational system to the base 1000, deduce criteria for divisibility by 37, 7, 11, 13.

2, a. Let $m > 0$, $(a, m) > 1$, let b be an integer, let x run through a complete, while ξ runs through a reduced, system of

49

residues modulo m. Prove that

$$\alpha) \quad \sum_x \left\{\frac{ax + b}{m}\right\} = \frac{1}{2}(m - 1),$$

$$\beta) \quad \sum_\xi \left\{\frac{a\xi}{m}\right\} = \frac{1}{2}\varphi(m).$$

b. Let $m > 0$, $(a, m) = 1$; let b, N, t be integers, $t > 0$; let $f(x) = \dfrac{ax + b}{m}$, $f(N) > 0$, $f(N + mt) > 0$. Prove, for the trapezoid bounded by the lines $x = N$, $x = N + mt$, $y = 0$, $y = f(x)$, that

(1) $$S = \sum \delta$$

where S is the area of the trapezoid, while the sum on the right is extended over all the lattice points of the trapezoid where $\delta = 1$ for the interior points, $\delta = \dfrac{1}{4}$ for the vertices, $\delta = \dfrac{1}{2}$ for the remaining points of the contour.

c. Letting, in contradistinction to problem **b**, $\delta = \dfrac{1}{6}$ for the vertices, prove formula (1) for a triangle with lattice point vertices.

3, a. Let $m > 0$, $(a, m) = 1$, $h \geqslant 0$, let c be a real number, let

$$S = \sum_{x=0}^{m-1} \left\{\frac{ax + \psi(x)}{m}\right\}$$

where $\psi(x)$ takes on values such that $c \leqslant \psi(x) \leqslant c + h$ for the values of x considered in the sum. Prove that

$$\left| S - \frac{1}{2}m \right| \leqslant h + \frac{1}{2}.$$

b. Let M be an integer, $m > 0$, $(a, m) = 1$, let A and B be real numbers, let

$$A = \frac{a}{m} + \frac{\lambda}{m^2}, \quad S = \sum_{x=M}^{M+m-1} \{Ax + B\}.$$

Prove that

$$\left| S - \frac{1}{2}m \right| \leqslant |\lambda| + \frac{1}{2}.$$

c. Let M be an integer, $m > 0$, $(a, m) = 1$,

$$S = \sum_{x=M}^{M+m-1} \{f(x)\},$$

where the function $f(x)$ has continuous derivatives $f'(x)$ and $f''(x)$ in the interval $M \leqslant x \leqslant M + m - 1$, while

$$f'(M) = \frac{a}{m} + \frac{\theta}{m^2}; \quad (a, m) = 1; \quad |\theta| < 1; \quad \frac{1}{A} \leqslant |f''(x)| \leqslant \frac{k}{A},$$

where

$$1 \leqslant m \leqslant \tau, \quad \tau = A^3, \quad A \geqslant 2, \quad k \geqslant 1.$$

Prove that

$$\left| S - \frac{1}{2}m \right| < \frac{k + 3}{2}$$

4. Let all the partial quotients in the continued fraction development of the irrational number A be bounded, let M be an integer, let m be a positive integer, and let B be a real number. Prove that

$$\sum_{x=M}^{M+m-1} \{Ax + B\} = \frac{1}{2}m + O(\ln m).$$

5, a. Let $A > 2$, $k \geqslant 1$ and let the function $f(x)$ have a continuous second derivative satisfying the condition

$$\frac{1}{A} \leqslant |f''(x)| \leqslant \frac{k}{A}$$

on the interval $Q \leqslant x \leqslant R$. Prove that

$$\sum_{Q<x\leqslant R} \{f(x)\} = \frac{1}{2}(R - Q) + \theta\Delta; \quad |\theta| < 1,$$

$$\Delta = (2k^2(R - Q) \ln A + 8kA)A^{-\frac{1}{3}}.$$

b. Let Q and R be integers, and let $0 < \sigma \leqslant 1$. Under the assumptions of problem **a**, prove that the number $\psi(\sigma)$ of fractions $\{f(x)\}$; $x = Q + 1, \ldots, R$ such that $0 \leqslant \{f(\sigma)\} < \sigma$ is given by the formula

$$\psi(\sigma) = \sigma(R - Q) + \theta' \cdot 2\Delta; \quad |\theta'| < 1.$$

6, a. Let T be the number of lattice points (x, y) of the region $x^2 + y^2 \leqslant r^2$ $(r \geqslant 2)$. Prove that

$$T = \pi r^2 + O(r^{\frac{2}{3}} \ln r).$$

b. Let n be an integer, $n > 2$, and let E be Euler's constant. Prove that

$$r(1) + r(2) + \ldots + r(n) = n(\ln n + 2E - 1) + O(n^{\frac{1}{3}} (\ln n)^2).$$

52

7. A system of n positive integers, each of which is represented to the base 2, is said to be proper if for every non-negative integer s, the number of integers in whose representation 2^s occurs, is even, and is said to be improper if this number is odd for at least one s.

Prove that an improper system can be made proper by decreasing or completely deleting some one of its members, while a proper system can be made improper by decreasing or completely deleting any one of its members.

8. a. Prove that the form

$$3^n x_n + 3^{n-1} x_{n-1} + \ldots + 3x_1 + x_0,$$

where $x_n, x_{n-1}, \ldots, x_1, x_0$ run through the values -1, 0, 1 independently of one another, represents the numbers

$$-H, \ldots, -1, 0, 1, \ldots, H; \quad H = \frac{3^{n+1} - 1}{3 - 1}$$

and represents each of them uniquely.

b. Let $m_1, m_2, \ldots, m_k$ be positive integers which are relatively prime in pairs. Using **c, §4**, prove that we obtain a complete residue system modulo $m_1 m_2 \ldots m_n$, by inserting in the form

$$x_1 + m_1 x_2 + m_1 m_2 x_3 + \ldots + m_1 m_2 \ldots m_{k-1} x_k$$

the numbers $x_1, x_2, \ldots, x_k$ which run through complete residue systems modulo $m_1, m_2, \ldots, m_k$.

9. Let $m_1, m_2, \ldots, m_k$ be integers which are relatively prime in pairs, and let

$$m_1 m_2 \ldots m_k = m_1 M_1 = m_2 M_2 = \ldots = m_k M_k.$$

a. Applying **c, §4**, prove that we obtain a complete system modulo $m_1 m_2 \ldots m_k$ by inserting in the form

$$M_1 x_1 + M_2 x_2 + \ldots + M_k x_k$$

53

the numbers x_1, x_2, ..., x_k which run through a complete system of residues modulo m_1, m_2, ..., m_k.

b. Applying **c, §4, ch. II** and **b, §5**, prove that we obtain a reduced system of residues modulo $m_1 m_2 \ldots m_k$ by inserting in the form

$$M_1 x_1 + M_2 x_2 + \ldots + M_k x_k$$

the numbers x_1, x_2, ..., x_k which run through a reduced residue system modulo m_1, m_2, ..., m_k.

c. Prove the theorem of problem **b** independently of theorem **c, §4, ch. II,** and then deduce the latter theorem from the former one.

d. Find an expression for $\varphi(p^a)$ by an elementary method, and using the equation in **c, §4, ch. II,** deduce an expression for $\varphi(a)$.

10. Let m_1, m_2, ..., m_k be integers greater than 1, which are relatively prime in pairs, and let $m = m_1 m_2 \ldots m_k$, $m_s M_s = m$.

a. Let x_1, x_2, ..., x_k, x run through complete residue systems, while ξ_1, ξ_2, ..., ξ_k, ξ run through reduced residue systems modulo m_1, m_2, ..., m_k, m. Prove that the fractions

$$\left\{ \frac{x_1}{m_1} + \frac{x_2}{m_2} + \ldots + \frac{x_k}{m_k} \right\}$$

coincide with the fractions $\left\{ \dfrac{x}{m} \right\}$, while the fractions

$\left\{ \dfrac{\xi_1}{m_1} + \dfrac{\xi_2}{m_2} + \ldots + \dfrac{\xi_k}{m_k} \right\}$ coincide with the fractions $\left\{ \dfrac{\xi}{m} \right\}$.

b. Consider k entire rational functions with integral coefficients of the r variables x, ..., w $(r \geqslant 1)$:

$$f_s(x, \ldots, w) = \sum_{a, \ldots, \delta} c_{a, \ldots, \delta}^{(s)} x^a \ldots w^\delta; \quad s = 1, \ldots, k,$$

54

and let

$$f(x, \ldots, w) = \sum_{a, \ldots, \delta} c_{a, \ldots, \delta} x^a \ldots w^\delta; \quad c_{a, \ldots, \delta} =$$

$$= \sum_{s=1}^{k} M_s c_{a, \ldots, \delta}^{(s)};$$

$x_s, \ldots, w_s$ run through complete residue systems, while $\xi_s, \ldots, \omega_s$ run through reduced residue systems modulo m_s; $x, \ldots, w$ run through complete residue systems, while $\xi, \ldots, \omega$ run through reduced residue systems modulo m. Prove that the fractions

$$\left\{ \frac{f_1(x_1, \ldots, w_1)}{m_1} + \ldots + \frac{f_k(x_k, \ldots, w_k)}{m_k} \right\}$$

coincide with the fractions $\left\{ \dfrac{f(x, \ldots, w)}{m} \right\}$, while the fractions

$$\left\{ \frac{f_1(\xi_1, \ldots, \omega_1)}{m_1} + \ldots + \frac{f_k(\xi_k, \ldots, \omega_k)}{m_k} \right\}$$

coincide with the fractions $\left\{ \dfrac{f(\xi, \ldots, \omega)}{m} \right\}$ (a generalization of the theorem of problem **a**).

11, a. Let m be a positive integer, let a be an integer, and let x run through a complete residue system modulo m. Prove that

$$\sum e^{2\pi i \frac{ax}{m}} = \begin{cases} m, \text{ if } a \text{ is a multiple of } m \\ \\ 0, \text{ otherwise.} \end{cases}$$

b. Let α be a real number, and let M and P be integers with $P > 0$. Letting (α) denote the numerical value of the difference between α and the integer closest to α (the distance

55

from α to the nearest integer), prove that

$$\left| \sum_{x=M}^{M+P-1} e^{2\pi i a x} \right| \leqslant \min \left(P, \frac{1}{h(\alpha)} \right); \quad h \geqslant \begin{cases} 2 \text{ always} \\ \\ 3, \text{ for } (\alpha) \leqslant \dfrac{1}{6}. \end{cases}$$

c. Let m be an integer, $m > 1$, and let the functions $M(a)$ and $P(a)$ take on integral values such that $P(a) > 0$ for the values $a = 1, 2, \ldots, m - 1$. Prove that

$$\sum_{a=1}^{m-1} \left| \sum_{x=M(a)}^{M(a)+P(a)-1} e^{2\pi i \frac{a}{m} x} \right| < \begin{cases} m \ln m - \dfrac{m}{3} \ln \left(2 \left[\dfrac{m}{6} \right] + 1 \right) \text{for } m \geqslant 6 \\ \\ m \ln m - \dfrac{m}{2}, \text{ for } m \geqslant 12, \\ \\ m \ln m - m, \text{ for } m \geqslant 60. \end{cases}$$

12, a. Let m be a positive integer, and let ξ run through a reduced residue system modulo m. Prove that

$$\mu(m) = \sum_{\xi} e^{2\pi i \frac{\xi}{m}}$$

b. Using the theorem of problem a, prove the first of the theorems of c, §3, ch. II (cf. solution of problem 28, a, ch. II).

c. Deduce the theorem of problem a, using the theorem of problem 17, a, ch. II.

d. Let

$$f(x, \ldots, w) = \sum_{a, \ldots, \delta} c_{a, \ldots, \delta} x^a \ldots w^\delta$$

be an entire rational function with integral coefficients of the r variables $x, \ldots, w (r \geqslant 1)$ and let a, m be integers with $m > 0$; $x, \ldots, w$ run through complete residue systems, while $\xi, \ldots, \omega$ run through reduced residue systems modulo m. We introduce the symbols

56

$$S_{a,m} = \sum_x \cdots \sum_w e^{2\pi i \frac{af(x,\ldots,w)}{m}},$$

$$S'_{a,m} = \sum_\xi \cdots \sum_\omega \exp(af(\xi,\ldots,\omega)/m)$$

Moreover, let $m = m_1 m_2 \ldots m_k$, where $m_1, \ldots, m_k$ are integers exceeding 1 which are relatively prime in pairs, and let $m_s M_s = m$. Prove that

$$S_{a_1,m_1} \ldots S_{a_k,m_k} = S_{M_1 a_1 + \ldots + M_k a_k, m},$$

$$S'_{a_1,m_1} \ldots S'_{a_k,m_k} = S'_{M_1 a_1 + \ldots + M_k a_k, m}.$$

e. Using the notation of problem **d** we set

$$A(m) = m^{-r} \sum_a S_{a,m}, \quad A'(m) = m^{-r} \sum_a S'_{a,m},$$

where a runs through a reduced residue system modulo m. Prove that

$$A(m_1) \ldots A(m_k) = A(m), \quad A'(m_1) \ldots A'(m_k) = A'(m).$$

13, a. Prove that

$$\varphi(a) = \sum_{n=0}^{a-1} \prod_p \left(1 - \frac{1}{p} \sum_{x=0}^{n-1} e^{2\pi i \frac{nx}{p}} \right)$$

where p runs through the prime divisors of the number a.

b. Deduce the well-known expression for $\varphi(a)$ from the identity of problem **a**.

14. Prove that

$$r(a) = \lim_{\epsilon \to 0} 2\epsilon \sum_{0 < x \leqslant \sqrt{a}} \sum_{k=1}^{\infty} k^{-(1+\epsilon)} \exp(2\pi i a k/x) + \delta$$

57

where $\delta = 1$ or $\delta = 0$, according as a is or is not the square of an integer.

15, a. Let p be a prime and let $h_1, h_2, \ldots, h_a$ be integers. Prove that

$$(h_1 + h_2 + \ldots + h_a)^p \equiv h_1^p + h_2^p + \ldots + h_a^p \pmod{p}.$$

b. Deduce Fermat's theorem from the theorem of problem **a.**

c. Deduce Euler's theorem from Fermat's theorem.

Numerical Exercises for Chapter III.

1, a. Find the remainder resulting from the division of $(12\,371^{56} + 34)^{28}$ by 111.

b. Is the number $2^{1093} - 2$ divisible by $1\,093^2$?

2, a. Applying the divisibility criteria of problem **1**, find the canonical decomposition of the number 244 943 325.

b. Find the canonical decomposition of the number 282 321 246 671 737.

CHAPTER IV

CONGRUENCES IN ONE UNKNOWN

§1. *Basic Concepts*

Our immediate problem is the study of congruences of the general form:

$$(1) \quad f(x) \equiv 0 (\mathrm{mod}\ m); \quad f(x) = ax^n + a_1 x^{n-1} + \ldots + a_n.$$

If a is not divisible by m, then n is said to be the *degree of the congruence.*

Solving a congruence means finding all the values of x which satisfy it. Two congruences which are satisfied by the same values of x are said to be *equivalent*.

If the congruence (1) is satisfied by some $x = x_1$, then (d, §2, ch. III) this congruence will also be satisfied by all numbers which are congruent to x_1 modulo m: $x \equiv x_1 \pmod{m}$. This whole class of numbers is considered to be *one solution*. In accordance with this convention, *congruence (1) has as many solutions as residues of a complete system satisfying it.*

Example. The congruence

$$x^5 + x + 1 \equiv 0 (\mathrm{mod}\ 7)$$

is satisfied by two numbers $x = 2$ and $x = 4$ among the numbers 0, 1, 2, 3, 4, 5, 6 of a complete residue system modulo 7.

59

Therefore the above congruence has the two solutions:

$$x \equiv 2(\text{mod } 7), \ x \equiv 4(\text{mod } 7).$$

§2. *Congruences of the First Degree*

a. A congruence of the first degree whose constant term has been placed on the right side (with opposite sign) can be put in the form

(1) $$ax \equiv b(\text{mod } m).$$

b. Turning to the investigation of the number of solutions, we first restrict the congruence by the condition $(a, m) = 1$. According to 1, our congruence has as many solutions as residues of a complete system satisfy it. But when x runs through a complete system of residues modulo m, ax also runs through a complete residue system (**d, 4, ch. III**). Therefore, in particular, ax will be congruent to b for one and only one value of x taken from the complete residue system. Therefore congruence (1) has one solution for $(a, m) = 1$.

c. Now let $(a, m) = d > 1$. Then, in order that the congruence (1) have a solution it is necessary (**e, $3, ch. III**) that b be divisible by d, for otherwise the congruence (1) is impossible for all integers x. Assuming then that b is a multiple of d, we set $a = a_1 d$, $b = b_1 d$, $m = m_1 d$. Then the congruence (1) is equivalent to the following one (obtained by dividing through by d): $a_1 x \equiv b_1(\text{mod } m_1)$, in which $(a_1, m_1) = 1$, and therefore it will have one solution modulo m_1. Let x_1 be the least non-negative residue of this solution modulo m_1, then all the numbers x which are solutions of this equation are found to be of the form

(2) $$x \equiv x_1(\text{mod } m_1).$$

But modulo m the numbers of (2) do not form one solution, but many solutions, and indeed as many solutions as there are numbers of (2) in the sequence $0, 1, 2, \ldots, m - 1$ of least non-negative residues modulo m. But these consist of the following numbers of (2):

$$x_1, \; x_1 + m_1, \; x_1 + 2m_1, \; \ldots, \; x_1 + (d - 1)m_1,$$

i.e. d numbers of the form (2), and hence the congruence (1) has d solutions.

d. Gathering together our results, we obtain the following theorem:

Let $(a, m) = d$. The congruence $ax \equiv b(\mathrm{mod}\ m)$ is impossible if b is not divisible by d. For b a multiple of d, the congruence has d solutions.

e. Turning to the finding of solutions of the congruence (1), we shall only consider a method which is based on the theory of continued fractions, where it is sufficient to restrict ourselves to the case in which $(a, m) = 1$.

Developing the fraction m/a in a continued fraction,

$$\frac{m}{a} = q_1 + \cfrac{1}{q_2 + \cfrac{1}{q_3 + \cfrac{\ddots}{ + \cfrac{1}{q_n}}}}$$

and considering the last two convergents:

$$\frac{P_{n-1}}{Q_{n-1}}, \; \frac{P_n}{Q_n} = \frac{m}{a}$$

by the properties of continued fractions (**e, §4, ch. I**) we have

$$mQ_{n-1} - aP_{n-1} = (-1)^n,$$

$$aP_{n-1} \equiv (-1)^{n-1}\ (\mathrm{mod}\ m),$$

$$a \cdot (-1)^{n-1} P_{n-1} b \equiv b(\mathrm{mod}\ m).$$

61

Hence, our congruence has the solution

$$x \equiv (-1)^{n-1} P_{n-1} b \pmod{m},$$

for whose calculation it is sufficient to calculate P_{n-1} by the method described in **d**, **§4, ch. I.**

Example. We solve the congruence

(3) $$111x \equiv 75 \pmod{321}.$$

Here $(111, 321) = 3$, while 75 is a multiple of 3. Therefore the congruence has three solutions.

Dividing both sides of the congruence and the modulus by 3, we obtain the congruence

(4) $$37x \equiv 25 \pmod{107},$$

which we must first solve. We have

```
         107 | 37
          74 |  2
      37 | 33
      33 |  1
  33 | 4
  32 | 8
4 | 1
4 | 4
  ''
```

q		2	1	8	4
P_s	1	2	3	26	107

Hence $n = 4$, $P_{n-1} = 26$, $b = 25$, and we have the solution of congruence (4) in the form

$$x \equiv -26 \cdot 25 \equiv 99 \pmod{107}.$$

62

From this the solutions of congruence (3) can be represented in the form:

$$x \equiv 99, \ 99 + 107, \ 99 + 2 \cdot 107 (\text{mod } 321),$$

i.e.

$$x \equiv 99, \ 206, \ 313 (\text{mod } 321).$$

§3. *Systems of Congruences of the First Degree*

a. We shall only consider the simplest system of congruences

(1) $x \equiv b_1 (\text{mod } m_1), \ x \equiv b_2 (\text{mod } m_2), \ \ldots, \ x \equiv b_k (\text{mod } m_k)$

in one unknown, but with different and pairwise prime moduli.

b. It is possible to solve the system (1), i.e. find all values of x satisfying it, by applying the following theorem:

Let the numbers M_s and M_s' be defined by the conditions

$$m_1 m_2 \ldots m_s = M_s m_s, \ M_s M_s' \equiv 1 (\text{mod } m_s)$$

and let

$$x_0 = M_1 M_1' b_1 + M_2 M_2' b_2 + \ldots + M_k M_k' b_k.$$

Then the set of values of x satisfying the system (1) are defined by the congruence

(2) $\qquad\qquad x \equiv x_0 (\text{mod } m_1 m_2 \ldots m_k)$

Indeed, in view of the fact that all the M_j which are different from M_s are divisible by m_s, for any $s = 1, 2, \ldots, k$, we have

$$x_0 \equiv M_s M_s' b_s \equiv b_s (\text{mod } m_s),$$

and therefore system (1) is satisfied by $x = x_0$. It follows immediately from this, that the system (1) is equivalent to

63

the system

(3) $x \equiv x_0 \pmod{m_1}$, $x \equiv x_0 \pmod{m_2}$, $\ldots$, $x \equiv x_0 \pmod{m_k}$

(i.e. the systems (1) and (3) are satisfied by the same values
of x). But the system (3), in view of the theorems of **c, §3,
ch. III** and **d, §3, ch. III**, is satisfied by those and only those
values of x which satisfy the congruence (2).

   c. *If b_1, b_2, $\ldots$, b_k independently run through complete
residue systems modulo m_1, m_2, $\ldots$, m_k, then x_0 runs through
a complete residue system modulo $m_1 m_2 \ldots m_k$.*

   Indeed, x_0 runs through $m_1 m_2 \ldots m_k$ values which are incongruent modulo $m_1 m_2 \ldots m_k$, in view of **d, §3, ch. III.**

   d. *Example.* We solve the system

$$x \equiv b_1 \pmod 4, \quad x \equiv b_2 \pmod 5, \quad x \equiv b_3 \pmod 7.$$

Here $4 \cdot 5 \cdot 7 = 4 \cdot 35 = 5 \cdot 28 = 7 \cdot 20$, while

$$35 \cdot 3 \equiv 1 \pmod 4, \quad 28 \cdot 2 \equiv 1 \pmod 5, \quad 20 \cdot 6 \equiv 1 \pmod 7.$$

Therefore

$$x = 35 \cdot 3b_1 + 28 \cdot 2b_2 + 20 \cdot 6b_3 = 105 b_1 + 56 b_2 + 120 b_3$$

and hence the set of values of x satisfying the system, can be
represented in the form

$$x \equiv 105 b_1 + 56 b_2 + 120 b_3 \pmod{140}.$$

   Thus, for example, the set of values satisfying the system

$$x \equiv 1 \pmod 4, \quad x \equiv 3 \pmod 5, \quad x \equiv 2 \pmod 7,$$

is

$$x \equiv 105 \cdot 1 + 56 \cdot 3 + 120 \cdot 2 \equiv 93 \pmod{140}$$

while the set of values of x satisfying the system

$$x \equiv 3 \,(\mathrm{mod}\ 4), \quad x \equiv 2 \,(\mathrm{mod}\ 5), \quad x \equiv 6 \,(\mathrm{mod}\ 7),$$

is

$$x \equiv 105 \cdot 3 + 56 \cdot 2 + 120 \cdot 6 \equiv 27 \,(\mathrm{mod}\ 140).$$

§4. *Congruences of Arbitrary Degree with Prime Modulus*

a. Let p be a prime. We shall prove general theorems relating to congruences of the form

(1) $\quad f(x) \equiv 0 \,(\mathrm{mod}\ p); \quad f(x) = ax^n + a_1 x^{n-1} + \ldots + a_n.$

b. *A congruence of the form* (1) *is equivalent to a congruence of degree not higher than* $p - 1$.

Indeed, dividing $f(x)$ by $x^p - x$, we have

$$f(x) = (x^p - x)Q(x) + R(x),$$

where the degree of $R(x)$ is not higher than $p - 1$. But $x^p - x \equiv 0 \,(\mathrm{mod}\ p)$ implies that $f(x) \equiv R(x) \,(\mathrm{mod}\ p)$, from which our theorem follows.

c. *If the congruence* (1) *has more than n solutions, then all the coefficients of* $f(x)$ *are multiples of* p.

Indeed, let the congruence (1) have at least $n + 1$ solutions. Letting $x_1, x_2, \ldots, x_n, x_{n+1}$ be the residues of these solutions, we can represent $f(x)$ in the form

$$
\begin{aligned}
(2) \quad f(x) =\ & a(x - x_1)(x - x_2)\ldots(x - x_{n-2})(x - x_{n-1})(x - x_n) + \\
& + b(x - x_1)(x - x_2)\ldots(x - x_{n-2})(x - x_{n-1}) + \\
& + c(x - x_1)(x - x_2)\ldots(x - x_{n-2}) + \\
& + \ldots\ldots\ldots\ldots\ldots\ldots\ldots\ldots\ldots\ldots\ldots + \\
& + k(x - x_1)(x - x_2) + \\
& + l(x - x_1) + \\
& + m.
\end{aligned}
$$

65

To this end, develop the summands on the right side into polynomials, and then choose b so that the sum of the coefficients of x^{n-1} in the first two polynomials coincide with a_1; knowing b, we choose c so that the sum of the coefficients of x^{n-2} in the first three polynomials coincides with a_2, etc.

Putting $x = x_1, x_2, \ldots, x_n, x_{n+1}$ successively in (2), we find that all the numbers $m, l, k, \ldots, c, b, a$ are multiples of p. This means that all the coefficients $a, a_1 \ldots, a_n$ are multiples of p (since they are sums of numbers which are multiples of p).

   d. *For prime p, we have the congruence (Wilson's theorem)*

(3) $$1 \cdot 2 \ldots (p - 1) + 1 \equiv 0 \,(\mathrm{mod}\, p).$$

Indeed, if $p = 2$, then the theorem is evident. If $p > 2$, then we consider the congruence

$$(x - 1)(x - 2)\ldots(x - (p - 1)) - (x^{p-1} - 1) \equiv 0 \,(\mathrm{mod}\, p);$$

its degree is not higher than $p - 2$ and it has $p - 1$ solutions, indeed solutions with residues $1, 2, \ldots, p - 1$. Therefore, by theorem c, all its coefficients are multiples of p; in particular the constant term is also divisible by p and the constant term is just equal to the left side of the congruence (3).

   Example. We have $1 \cdot 2 \cdot 3 \cdot 4 \cdot 5 \cdot 6 + 1 = 721 \equiv 0 \,(\mathrm{mod}\, 7)$.

§5. *Congruences of Arbitrary Degree with Composite Modulus*

   a. *If $m_1, m_2, \ldots, m_k$ are pairwise prime, then the congruence*

(1) $$f(x) \equiv 0 \,(\mathrm{mod}\, m_1 m_2 \ldots m_k)$$

is equivalent to the system

$$f(x) \equiv 0 \,(\mathrm{mod}\, m_1),$$

$$f(x) \equiv 0 \,(\mathrm{mod}\, m_2), \ldots, f(x) \equiv 0 \,(\mathrm{mod}\, m_k).$$

Letting T_1, T_2, ..., T_k be the numbers of solutions of the individual congruences of this system with respect to the corresponding moduli, and letting T be the number of solutions of the congruence (1), *we have*

$$T = T_1 T_2 \ldots T_k.$$

Indeed, the first part of the theorem follows from **c** and **d, §3, ch. III.** The second part of the theorem follows from the fact that each congruence

(2) $$f(x) \equiv 0 \,(\mathrm{mod}\ m_s)$$

is satisfied if and only if one of the T_s congruences of the form

$$x \equiv b_s \,(\mathrm{mod}\ m_s),$$

where b_s runs through the residues of the solutions of the congruence (2), is satisfied, while all $T_1 T_2 \ldots T_k$ different combinations of the form

$$x \equiv b_1 \,(\mathrm{mod}\ m_1), \ x \equiv b_2 \,(\mathrm{mod}\ m_2), \ \ldots, \ x \equiv b_k \,(\mathrm{mod}\ m_k),$$

are possible, which leads (**c, §3**) to different classes modulo $m_1 m_2 \ldots m_k$.

Example. The congruence

(3) $$f(x) \equiv 0 \,(\mathrm{mod}\ 35), \ f(x) = x^4 + 2x^3 + 8x + 9$$

is equivalent to the system

$$f(x) \equiv 0 \,(\mathrm{mod}\ 5), \ f(x) \equiv 0 \,(\mathrm{mod}\ 7).$$

It is easy (**§1**) to verify that the first congruence of this system has two solutions: $x \equiv 1; 4 \,(\mathrm{mod}\ 5)$, the second congruence has three solutions: $x \equiv 3; 5; 6 \,(\mathrm{mod}\ 7)$. Hence the congruence (3) has $2 \cdot 3 = 6$ solutions. In order to find these six solutions, we must solve six systems of the form

(4) $\qquad x \equiv b_1 \,(\text{mod } 5), \; x \equiv b_2 \,(\text{mod } 7),$

which we obtain by letting b_1 run through the values $b_1 \equiv 1;\ 4,$ while b_2 runs through the values $b_2 \equiv 3;\ 5;\ 6.$ But since

$$35 = 5 \cdot 7 = 7 \cdot 5, \; 7 \cdot 3 \equiv 1\,(\text{mod } 5), \; 5 \cdot 3 \equiv 1\,(\text{mod } 7),$$

the set of values of x satisfying the system (4) can be represented in the form (b, §3)

$$x \equiv 21 b_1 + 15 b_2 \,(\text{mod } 35).$$

Therefore the solutions of congruence (3) are

$$x \equiv 31;\ 26;\ 6;\ 24;\ 19;\ 34 \,(\text{mod } 35).$$

b. In view of theorem **a** the investigation and solution of congruences of the form

$$f(x) \equiv 0 \,(\text{mod } p_1^{\alpha_1} p_2^{\alpha_2} \ldots p_k^{\alpha_k})$$

reduces to the investigation and solution of congruences of the form

(5) $\qquad f(x) \equiv 0 \,(\text{mod } p^{\alpha});$

this last congruence reduces in general, as we shall soon see, to the congruence

(6) $\qquad f(x) \equiv 0 \,(\text{mod } p)$

Indeed, every x satisfying the congruence (5) must necessarily satisfy the congruence (6). Let

$$x \equiv x_1 \,(\text{mod } p)$$

be any solution of the congruence (6). Then $x = x_1 + pt_1$, where t_1 is an integer. Inserting this value of x in the congruence

$$f(x) \equiv 0 \, (\text{mod } p^2)$$

and developing the left side by means of the Taylor formula, we find (noting that $\dfrac{1}{k!} f^{(k)}(x_1)$ is an integer, and deleting the terms which are multiples of p^2)

$$f(x_1) + pt_1 f'(x_1) \equiv 0 (\text{mod } p^2), \quad \frac{f(x_1)}{p} + t_1 f'(x_1) \equiv 0 (\text{mod } p).$$

Restricting ourselves to the case in which $f'(x_1)$ is not divisible by p, we have one solution:

$$t_1 \equiv t_1' \, (\text{mod } p); \quad t_1 = t_1' + pt_2.$$

The expression for x takes on the form

$$x = x_1 + p^2 t_1' + p^2 t_2 = x_2 + p^2 t_2;$$

inserting it in the congruence

$$f(x) \equiv 0 \, (\text{mod } p^3),$$

we find

$$f(x_2) + p^2 t_2 f'(x_2) \equiv 0 \, (\text{mod } p^3)$$

$$\frac{f(x_2)}{p^2} + t_2 f'(x_2) \equiv 0 \, (\text{mod } p).$$

Here $f'(x_2)$ is not divisible by p since

$$x_2 \equiv x_1 \, (\text{mod } p),$$

$$f'(x_2) \equiv f'(x_1) \, (\text{mod } p),$$

69

and hence the latter equation has one solution:

$$t_2 \equiv t_2' \,(\text{mod } p),$$

$$t_2 = t_2' + p t_3.$$

The expression for x takes on the form

$$x = x_2 + p^2 t_2' + p^3 t_3 = x_3 + p^3 t_3;$$

and so forth. In this way, given a solution of the congruence (6) we can find a solution of the congruence (5) which is congruent to it. *Hence, if $f'(x_1)$ is not divisible by p, each solution $x \equiv x_1 \,(\text{mod } p)$ of the congruence (6) gives a solution of the congruence* (5):

$$x = x_\alpha + p^\alpha t_\alpha;$$

$$x \equiv x_\alpha \,(\text{mod } p^\alpha).$$

Example. We solve the congruence

(7)
$$\begin{cases} f(x) \equiv 0 \,(\text{mod } 27); \\ f(x) = x^4 + 7x + 4. \end{cases}$$

The congruence $f(x) \equiv 0 \,(\text{mod } 3)$ has one solution $x \equiv 1 \,(\text{mod } 3)$; here $f'(1) \equiv 2 (\text{mod } 3)$, and hence, is not divisible by 3. We find

$$x = 1 + 3t_1,$$

$$f(1) + 3t_1 f'(1) \equiv 0 \,(\text{mod } 9), \quad 3 + 3t_1 \cdot 2 \equiv 0 \,(\text{mod } 9),$$

$$2t_1 + 1 \equiv 0 \,(\text{mod } 3), \quad t_1 \equiv 1 \,(\text{mod } 3), \quad t_1 = 1 + 3t_2,$$

$$x = 4 + 9t_2,$$

$$f(4) + 9t_2 f'(4) \equiv 0 \,(\text{mod } 27), \quad 18 + 9t_2 \cdot 2 \equiv 0 \,(\text{mod } 27),$$

$$2t_2 + 2 \equiv 0 \,(\text{mod } 3), \quad t_2 \equiv 2 \,(\text{mod } 3), \quad t = 2 + 3t_3,$$

$$x = 22 + 27t_3.$$

70

Therefore, the congruence (7) has one solution:

$$x \equiv 22 \,(\text{mod } 27).$$

Problems for Chapter IV

1, a. Let m be a positive integer and let $f(x, \ldots, w)$ be an entire rational function with integral coefficients of the r variables $x, \ldots, w (r \geqslant 1)$. If the system $x = x_0, \ldots, w = w_0$ satisfies the congruence

(1) $$f(x, \ldots, w) \equiv 0 \,(\text{mod } m),$$

then (generalizing the definition of §1) the system of classes of integers modulo m:

$$x \equiv x_0 \,(\text{mod } m), \ \ldots, \ w \equiv w_0 \,(\text{mod } m)$$

will be considered to be one solution of the congruence (1).

Let T be the number of solutions of the congruence (1). Prove that

$$Tm = \sum_{a=0}^{m-1} \sum_{x=0}^{m-1} \cdots \sum_{w=0}^{m-1} e^{2\pi i \frac{af(x, \ldots, w)}{m}}$$

b. Using the notation of problem **a** and problem **12, e, ch. III**, prove that

$$Tm = m^r \sum_{m_0 \backslash m} A(m_0).$$

c. Apply the equation of problem **a** to the proof of the theorem on the number of solutions of a congruence of the first degree.

d. Let m be a positive integer; let $a, \ldots, f, g$ be $r + 1 \,(r > 0)$ integers; $d = (a, \ldots, f, m)$; let T be the

71

number of solutions of the congruence

$$ax + \ldots + fw + g \equiv 0 \,(\text{mod } m).$$

Using the equation of problem **a**, prove that

$$T = \begin{cases} m^{r-1}d, & \text{if } g \text{ is a multiple of } d, \\ 0, & \text{otherwise.} \end{cases}$$

e. Prove the theorem of problem **d**, starting from the theorem on the number of solutions of the congruence $ax \equiv b \,(\text{mod } m)$.

2, a. Let $m > 1$, $(a, m) = 1$. Prove that the congruence $ax \equiv b \,(\text{mod } m)$ has the solution $x \equiv ba^{\varphi(m)-1} \,(\text{mod } m)$.

b. Let p be a prime, $0 < a < p$. Prove that the congruence $ax \equiv b \,(\text{mod } p)$ has the solution

$$x \equiv b(-1)^{a-1} \frac{(p-1)(p-2)\ldots(p-a+1)}{1 \cdot 2 \ldots a} \,(\text{mod } p).$$

c, $\alpha)$ Find the simplest possible method of solving a congruence of the form

$$2^k x \equiv b \,(\text{mod } m); \quad (2, m) = 1.$$

$\beta)$ Find the simplest possible method of solving a congruence of the form

$$3^k x \equiv b \,(\text{mod } m); \quad (3, m) = 1.$$

$\gamma)$ Let $(a, m) = 1$, $1 < a < m$. Applying the methods used in problems $\alpha)$ and $\beta)$, prove that finding the solutions of the congruence $ax \equiv b \,(\text{mod } m)$ can be reduced to finding the solutions of a congruence of the form $b + mt \equiv 0 \,(\text{mod } p)$ where p is a prime divisor of the number a.

3. Let m be an integer, $m > 1$, $1 < r < m$, $(a, m) = 1$. Using the theory of congruences prove the existence of integers x and y such that

$$ax \equiv y \,(\text{mod } m), \; 0 < x \leqslant \tau, \; 0 < |y| < \frac{m}{\tau}.$$

4, a. For $(a, m) = 1$, we will consider the symbolic fraction $\dfrac{b}{a}$ modulo m, which denotes any residue of a solution of the congruence $ax \equiv b \,(\text{mod } m)$. Prove that (the congruences are taken modulo m)

$\alpha)$ For $a \equiv a_1$, $b \equiv b_1$ we have $\dfrac{b}{a} \equiv \dfrac{b_1}{a_1}$.

$\beta)$ The numerator b of the symbolic fraction $\dfrac{b}{a}$ can be replaced by a congruent b_0 which is a multiple of a. Then the symbolic fraction $\dfrac{b}{a}$ is congruent to the ordinary fraction $\dfrac{b_0}{a}$, where the congruence is taken with ordinary integers.

$\gamma) \; \dfrac{b}{a} + \dfrac{d}{c} \equiv \dfrac{bc + ad}{ac}$

$\delta) \; \dfrac{b}{a} \cdot \dfrac{d}{c} \equiv \dfrac{bd}{ac}$

b, $\alpha)$ Let p be a prime, $p > 2$, and let a be an integer, $0 < a < p - 1$. Prove that

$$\binom{p-1}{a} \equiv (-1)^a \,(\text{mod } p).$$

$\beta)$ Let p be a prime, $p > 2$. Prove that

$$\frac{2^p - 2}{p} \equiv 1 - \frac{1}{2} + \frac{1}{3} - \cdots - \frac{1}{p-1} \,(\text{mod } p).$$

73

5, a. Let d be a divisor of the number a which is not divisible by primes smaller than n, and let κ be the number of different divisors of the number d. Prove that the number of multiples of d in the sequence

$$(1)\ 1 \cdot 2 \ldots n, \ 2 \cdot 3 \ldots (n+1), \ \ldots, \ a(a+1) \ldots (a+n-1)$$

is $\dfrac{n^{\kappa} a}{d}$.

b. Let $p_1, p_2, \ldots, p_k$ be the different prime divisors of the number a which are not smaller than n. Prove that the number of integers of the sequence (1) relatively prime to a is

$$a \left(1 - \frac{n}{p_1} \right) \left(1 - \frac{n}{p_2} \right) \ldots \left(1 - \frac{n}{p_k} \right)$$

6. Let $m_{1,2,\ldots,k}$ be the least common multiple of the numbers $m_1, m_2, \ldots, m_k$.

a. Let $d = (m_1, m_2)$. Prove that the system

$$x \equiv b_1 \, (\mathrm{mod}\ m_1), \ x \equiv b_2 \, (\mathrm{mod}\ m_2)$$

is solvable if and only if $b_2 - b_1$ is a multiple of d, and if the system is solvable, the set of values of x satisfying this system is determined by a congruence of the form

$$x \equiv x_{1,2} \, (\mathrm{mod}\ m_{1,2}).$$

b. Prove that, if the system

$$x \equiv b_1 \, (\mathrm{mod}\ m_1), \ x \equiv b_2 \, (\mathrm{mod}\ m_2), \ \ldots, \ x \equiv b_k \, (\mathrm{mod}\ m_k)$$

is solvable, the set of values of x satisfying it is determined by a congruence of the form

$$x \equiv x_{1,2,\ldots,k} \, (\mathrm{mod}\ m_{1,2,\ldots,k}).$$

74

7. Let m be an integer, $m > 1$, let a and b be integers, and let

$$\left(\frac{a, b}{m}\right) = \sum_x \exp\left(2\pi i \frac{ax + bx'}{m}\right)$$

where x runs through a reduced residue system modulo m, while $x' \equiv \dfrac{1}{x} \pmod{m}$ (in the sense of problem **4, a**). Prove the following properties of the symbol $\left(\dfrac{a, b}{m}\right)$:

$\alpha)$ $\left(\dfrac{a, b}{m}\right)$ is real.

$\beta)$ $\left(\dfrac{a, b}{m}\right) = \left(\dfrac{b, a}{m}\right)$.

$\gamma)$ For $(h, m) = 1$, we have $\left(\dfrac{a, bh}{m}\right) = \left(\dfrac{ah, b}{m}\right)$.

$\delta)$ For $m_1, m_2, \ldots, m_k$ relatively prime in pairs, setting $m_1 m_2 \ldots m_k = m$, $m = m_s M_s$, we have

$$\left(\frac{a_1, 1}{m_1}\right) \left(\frac{a_2, 1}{m_2}\right) \cdots \left(\frac{a_k, 1}{m_k}\right) =$$

$$= \left(\frac{M_1^2 a_1 + M_2^2 a_2 + \ldots + M_k^2 a_k, 1}{m}\right).$$

8. Let the congruence

$$a_0 x^n + a_1 x^{n-1} + \ldots + a_n \equiv 0 \pmod{p}$$

have the n solutions

$$x \equiv x_1, x_2, \ldots, x_n \pmod{p}.$$

75

Prove that

$$a_1 \equiv -a_0 S_1 \pmod{p},$$

$$a_2 \equiv a_0 S_2 \pmod{p},$$

$$a_3 \equiv -a_0 S_3 \pmod{p},$$

$$\cdots\cdots\cdots\cdots\cdots$$

$$a_n \equiv (-1)^n a_0 S_n \pmod{p},$$

where S_1 is the sum of all the x_s, S_2 is the sum of the products of pairs of the x_s, S_3 is the sum of the products of triples of the x_s, etc.

9, a. Prove Wilson's theorem by considering pairs x, x' of numbers of the sequence $2, 3, \ldots, p - 2$, satisfying the condition $xx' \equiv 1 \pmod{p}$.

b. Let P be an integer, $P > 1, 1 \cdot 2 \ldots (P - 1) + 1 \equiv \equiv 0 \pmod{P}$. Prove that P is a prime.

10, a. Let $(a_0, m) = 1$. Find a congruence of degree $n(n > 0)$ with leading coefficient 1, equivalent to the congruence

$$a_0 x^n + a_1 x^{n-1} + \ldots + a_n \equiv 0 \pmod{m}.$$

b. Prove that a necessary and sufficient condition in order that the congruence $f(x) \equiv 0 \pmod{p}$; $f(x) = x^n + a_1 x^{n-1} + + \ldots + a_n$; $n \leqslant p$; has n solutions, is the divisibility by p of all the coefficients of the remainder after the division of $x^p - x$ by $f(x)$.

c. Let n be a divisor of $p - 1$; $n > 1$; $(A, p) = 1$. Prove that a necessary and sufficient condition for the solvability of the congruence $x^n \equiv A \pmod{p}$ is $A^{\frac{p-1}{n}} \equiv 1 \pmod{p}$, while if the congruence is solvable, it has n solutions.

11. Let n be a positive integer, $(A, m) = 1$, we assume that we know a solution $x \equiv x_0 \pmod{m}$ of the congruence

$x^n \equiv A \pmod{m}$. Prove that all the solutions of this congruence can be represented as the product of x_0 and a residue of a solution of the congruence $y^n \equiv 1 \pmod{m}$.

Numerical Exercises for Chapter IV

1, a. Solve the congruence $256x \equiv 179 \pmod{337}$.
   b. Solve the congruence $1215x \equiv 560 \pmod{2755}$.
   2, a. Solve the congruences of exercises 1, a and 1, b by the method of problem 2, c.
   b. Solve the congruence $1296x \equiv 1105 \pmod{2413}$ by the method of problem 2, c.
   3. Find all pairs x, y satisfying the indeterminate equation $1245x - 1603y = 999$.
   4, a. Find a general solution of the system

$$x \equiv b_1 \pmod{13}, \quad x \equiv b_2 \pmod{17}.$$

Using this general solution, find three numbers whose division by 13 and 17 gives the respective remainders 1 and 12, 6 and 8, 11 and 4.
   b. Find a general solution for the system

$$x \equiv b_1 \pmod{25}, \quad x \equiv b_2 \pmod{27}, \quad x \equiv b_3 \pmod{59}.$$

5, a. Solve the system of congruences

$$x \equiv 3 \pmod{8}, \quad x \equiv 11 \pmod{20}, \quad x \equiv 1 \pmod{15}.$$

b. Solve the system of congruences

$$x \equiv 1 \pmod{3}, \quad x \equiv 4 \pmod{5}, \quad x \equiv 2 \pmod{7},$$

$$x \equiv 9 \pmod{11}, \quad x \equiv 3 \pmod{13}.$$

6. Solve the system of congruences

$$x + 4y - 29 \equiv 0 \pmod{143}, \quad 2x - 9y + 84 \equiv 0 \pmod{143}.$$

7, a. What congruence of degree smaller than 5 is equivalent to the congruence

$$3x^{14} + 4x^{13} + 3x^{12} + 2x^{11} + x^9 + 2x^8 + 4x^7 + x^6 +$$

$$+ 3x^4 + x^3 + 4x^2 + 2x \equiv 0 \,(\mathrm{mod}\ 5)?$$

b. What congruence of degree smaller than 7 is equivalent to the congruence

$$2x^{17} + 6x^{16} + x^{14} + 5x^{12} + 3x^{11} + 2x^{10} + x^9 + 5x^8 +$$

$$+ 2x^7 + 3x^5 + 4x^4 + 6x^3 + 4x^2 + x + 4 \equiv 0 \,(\mathrm{mod}\ 7)?$$

8. What congruence with leading coefficient 1 is equivalent to the congruence (problem **10, a**)

$$70x^6 + 78x^5 + 25x^4 + 68x^3 + 52x^2 + 4x + 3 \equiv 0 \,(\mathrm{mod}\ 101)?$$

9, a. Solve the congruence

$$f(x) \equiv 0 \,(\mathrm{mod}\ 27), \quad f(x) = 7x^4 + 19x + 25,$$

by first finding all the solutions of the congruence

$$f(x) \equiv 0 \,(\mathrm{mod}\ 3)$$

by trial.

b. Solve the congruence $9x^2 + 29x + 62 \equiv 0 \,(\mathrm{mod}\ 64)$.

10, a. Solve the congruence $x^3 + 2x + 2 \equiv 0 \,(\mathrm{mod}\ 125)$.

b. Solve the congruence $x^4 + 4x^3 + 2x^2 + 2x + 12 \equiv$ $\equiv 0 \,(\mathrm{mod}\ 625)$.

11, a. Solve the congruence $6x^3 + 27x^2 + 17x + 20 \equiv$ $\equiv 0 \,(\mathrm{mod}\ 30)$.

b. Solve the congruence $31x^4 + 57x^3 + 96x + 191 \equiv$ $\equiv 0 \,(\mathrm{mod}\ 225)$.

CHAPTER V

CONGRUENCES OF THE SECOND DEGREE

§1. *General Theorems*

a. We shall only consider the simplest of the congruences of degree $n > 1$, i.e. *the two-term congruences*:

(1) $$x^n \equiv a \,(\mathrm{mod}\ m); \quad (a,\ m) = 1$$

If the congruence (1) has solutions, then a is said to be an *n-th power residue*, otherwise a is said to be an *n-th power non-residue*. In particular, for $n = 2$ the residues or non-residues are said to be *quadratic*, for $n = 3$ *cubic*, for $n = 4$ *biquadratic*.

In this chapter we shall consider the case $n = 2$ in detail and we first consider the two-term congruences of the second degree for odd prime modulus p:

(2) $$x^2 \equiv a \,(\mathrm{mod}\ p); \quad (a,\ p) = 1.$$

c. *If a is a quadratic residue modulo p, then the congruence* (2) *has two solutions.*

Indeed, if a is a quadratic residue, then the congruence (2) has at least one solution $x \equiv x_1 \,(\mathrm{mod}\ p)$. But since $(-x_1)^2 = x_1^2$, the same congruence also has the second solution $x \equiv -x_1 \,(\mathrm{mod}\ p)$. This second solution is different from the

79

first since $x_1 \equiv -x_1 \pmod{p}$ would imply $2x_1 \equiv 0 \pmod{p}$, which is impossible since $(2, p) = (x_1, p) = 1$.

These two solutions exhaust all the solutions of the congruence (2) since the latter, being a congruence of the second degree, cannot have more than two solutions (**c**, §**4, ch. IV**).

d. *A reduced residue system modulo p consists of* $\dfrac{p-1}{2}$

quadratic residues which are congruent to the numbers

$$(3) \qquad 1^2, \ 2^2, \ \ldots, \ \left(\frac{p-1}{2}\right)^2$$

and $\dfrac{p-1}{2}$ *quadratic non-residues.*

Indeed, among the residues of a reduced system modulo p, the quadratic residues are those and only those which are squares of the numbers (a reduced system of residues)

$$(4) \qquad -\frac{p-1}{2}, \ \ldots, \ -2, \ -1, \ 1, \ 2, \ \ldots, \ \frac{p-1}{2}$$

i.e. with the numbers of (3). Here the numbers of (3) are incongruent modulo p, since $k^2 \equiv l^2 \pmod{p}$, $0 < k < l \leqslant \dfrac{p-1}{2}$, it would follow that the congruence $x^2 \equiv l^2 \pmod{p}$ is satisfied by four numbers: $x = -l, \ -k, \ k, \ l$ among the numbers of (4), contradicting **c**.

e. *If a is a quadratic residue modulo p, then*

$$(5) \qquad a^{\frac{p-1}{2}} \equiv 1 \pmod{p};$$

if a is a quadratic non-residue modulo p, then

$$(6) \qquad a^{\frac{p-1}{2}} \equiv -1 \pmod{p}.$$

80

Indeed, by Fermat's theorem,

$$a^{p-1} \equiv 1 \,(\text{mod } p); \quad \left(a^{\frac{p-1}{2}} - 1 \right) \left(a^{\frac{p-1}{2}} + 1 \right) \equiv 0 \,(\text{mod } p).$$

One and only one of the factors of the left side of the latter congruence is divisible by p (both factors cannot be divisible by p, for if they were, then 2 would be divisible by p). Therefore one and only one of the congruences (5) and (6) can hold.

But every quadratic residue a satisfies the congruence

(7) $$a \equiv x^2 \,(\text{mod } p)$$

for some x, and therefore also satisfies the congruence (5), which can be obtained by raising each side of (7) to the power $\dfrac{p-1}{2}$. Here the quadratic residues exhaust all the solutions of the congruence (5), since it cannot have more than $\dfrac{p-1}{2}$ solutions because it is a congruence of degree $\dfrac{p-1}{2}$.

Therefore the quadratic non-residues satisfy the congruence (6).

§2. The Legendre Symbol

a. We now consider *Legendre's symbol* $\left(\dfrac{a}{p} \right)$ (read as: the symbol of a with respect to p). This symbol is defined for all a which are not divisible by p; it is equal to 1 if a is a quadratic residue, and equal to -1 if a is a quadratic non-residue. The number a is said to be the numerator, the number p the denominator, of the symbol.

b. In view of **e, §1**, it is evident that we have

$$\left(\frac{a}{p} \right) \equiv a^{\frac{p-1}{2}} \,(\text{mod } p).$$

81

c. Here we deduce the most important properties of the Legendre symbol and in the next paragraph, the properties of the generalization of this symbol—Jacobi's symbol, which is useful for the rapid calculation of this symbol, and hence solves the problem of the possibility of the congruence

$$x^2 \equiv a \,(\mathrm{mod}\ p).$$

d. *If* $a \equiv a_1$ (mod p), *then* $\left(\dfrac{a}{p}\right) = \left(\dfrac{a_1}{p}\right)$.

This property follows from the fact that the numbers of an equivalence class are all either quadratic residues or non-residues.

e. $\left(\dfrac{1}{p}\right) = 1.$

Indeed, $1 = 1^2$ and hence 1 is a quadratic residue.

f. $\left(\dfrac{-1}{p}\right) = (-1)^{\frac{p-1}{2}}.$

This property follows from **b** for $a = -1$.

Since $\dfrac{p-1}{2}$ is even for p of the form $4m + 1$ and odd for p of the form $4m + 3$, it follows that -1 is a quadratic residue of primes of the form $4m + 1$ and a quadratic non-residue of primes of the form $4m + 3$.

g. $\left(\dfrac{ab\ldots l}{p}\right) = \left(\dfrac{a}{p}\right)\left(\dfrac{b}{p}\right)\cdots\left(\dfrac{l}{p}\right).$

Indeed, we have

$$\left(\frac{ab\ldots l}{p}\right) \equiv (ab\ldots l)^{\frac{p-1}{2}} \equiv a^{\frac{p-1}{2}} b^{\frac{p-1}{2}} \ldots l^{\frac{p-1}{2}} \equiv$$

$$\equiv \left(\frac{a}{p}\right)\left(\frac{b}{p}\right)\cdots\left(\frac{l}{p}\right) \ (\mathrm{mod}\ p),$$

82

from which it follows that our assertion is true. A consequence of our result is

$$\left(\frac{ab^2}{p}\right) = \left(\frac{a}{p}\right)$$

i.e. we can delete any square factor from the numerator of a symbol.

h. In order to deduce further properties of Legendre's symbol, we first give it another interpretation. Setting $p_1 = \dfrac{p-1}{2}$, we consider the congruences

$$(1) \quad \begin{cases} a \cdot 1 \equiv \epsilon_1 r_1 \,(\text{mod } p) \\[2mm] a \cdot 2 \equiv \epsilon_2 r_2 \,(\text{mod } p) \\[2mm] \cdots\cdots\cdots\cdots\cdots\cdots\cdots\cdots\cdots\cdots\cdots \\[2mm] a \cdot p_1 \equiv \epsilon_{p_1} r_{p_1} \,(\text{mod } p); \; p_1 = \dfrac{p-1}{2}. \end{cases}$$

where $\epsilon_x r_x$ is the absolutely least residue of ax and r_x is its modulus so that $\epsilon_x = \pm 1$.

The numbers $a \cdot 1, -a \cdot 1, a \cdot 2, -a \cdot 2, \ldots, a \cdot p_1, -a \cdot p_1$ form a reduced residue system modulo p (**c, §5, ch. III**); their absolutely least residues are just $\epsilon_1 r_1, -\epsilon_1 r_1, \epsilon_2 r_2, -\epsilon_2 r_2, \ldots,$ $\epsilon_{p_1} r_{p_1}, -\epsilon_{p_1} r_{p_1}$. Those which are positive i.e. $r_1, r_2, \ldots, r_{p_1},$ must coincide with the numbers $1, 2, \ldots, p_1$ (**b, §4, ch. III**).

Multiplying together the congruences (1) and dividing through by

$$1 \cdot 2 \ldots p_1 = r_1 r_2 \ldots r_{p_1},$$

we find $a^{\frac{p-1}{2}} \equiv \epsilon_1 \epsilon_2 \ldots \epsilon_{p_1} \,(\text{mod } p)$, from which (**b**) we have

$$(2) \qquad\qquad \left(\frac{a}{p}\right) = \epsilon_1 \epsilon_2 \ldots \epsilon_{p_1}$$

83

i. The expression for Legendre's symbol which we have found can be put in a more concise form. We have

$$\left[\frac{2ax}{p}\right] = \left[2\left[\frac{ax}{p}\right] + 2\left\{\frac{ax}{p}\right\}\right] = 2\left[\frac{ax}{p}\right] + \left[2\left\{\frac{ax}{p}\right\}\right],$$

which is even or odd according as the least positive residue of the number ax is less or greater than $\frac{1}{2}p$, i.e. according as $\epsilon_x = 1$ or $\epsilon_x = -1$. It is evident from this that

$$\epsilon_x = (-1)^{\left[\frac{2ax}{p}\right]}$$

and therefore we find from (2) that

$$\left(\frac{a}{p}\right) = (-1)^{\sum\limits_{x=1}^{p_1}\left[\frac{2ax}{p}\right]}$$

j. Assuming a to be odd, we transform the latter equation. We have ($a + p$ is even)

$$\left(\frac{2a}{p}\right) = \left(\frac{2a + 2p}{p}\right) = \left(\frac{4\cdot\dfrac{a+p}{2}}{p}\right) = \left(\frac{\dfrac{a+p}{2}}{p}\right) =$$

$$= (-1)^{\sum\limits_{x=1}^{p_1}\left[\frac{(a+p)x}{p}\right]} = (-1)^{\sum\limits_{x=1}^{p_1}\left[\frac{ax}{p}\right] + \sum\limits_{x=1}^{p_1} x}$$

and hence

(3) $$\left(\frac{2}{p}\right)\left(\frac{a}{p}\right) = (-1)^{\sum\limits_{x=1}^{p_1}\left[\frac{ax}{p}\right] + \frac{p^2-1}{8}}$$

The formula (3) allows us to deduce two very important properties of the Legendre symbol.

84

k. $\left(\dfrac{2}{p}\right) = (-1)^{\frac{p^2-1}{8}}.$

This follows from formula (3) for $a = 1$.
Moreover, since

$$\frac{(8m \pm 1)^2 - 1}{8} = 8m^2 \pm 2m, \text{ even}$$

while

$$\frac{(8m \pm 3)^2 - 1}{8} = 8m^2 \pm 6m + 1, \text{ odd,}$$

it follows that 2 is a quadratic residue of primes of the form $8m \pm 1$ ($8m + 1$, $8m + 7$) and a quadratic non-residue of primes of the form $8m \pm 3$ ($8m + 3$, $8m + 5$).

l. *If* p *and* q *are odd primes, then* (*the quadratic reciprocity law*)

$$\left(\frac{q}{p}\right) = (-1)^{\frac{p-1}{2} \cdot \frac{q-1}{2}} \left(\frac{p}{q}\right)$$

Since $\dfrac{p-1}{2} \cdot \dfrac{q-1}{2}$ is odd only in the case in which both numbers p and q are of the form $4m + 3$ and even if one of these numbers is of the form $4m + 1$, the above property can be formulated as follows:

If both the numbers p and q are of the form $4m + 3$, then

$$\left(\frac{q}{p}\right) = -\left(\frac{p}{q}\right);$$

if one of them is of the form $4m + 1$, then

$$\left(\frac{q}{p}\right) = \left(\frac{p}{q}\right).$$

85

In order to prove our results, we note that, in view of **k**, formula (3) takes on the form

$$(4) \qquad \left(\frac{a}{p}\right) = (-1)^{\sum\limits_{x=1}^{p_1} \left[\frac{ax}{p}\right]}$$

Setting $\dfrac{q-1}{2} = q_1$, we consider $p_1 q_1$ pairs of numbers which are obtained when the numbers a and y in the expressions qx, py run through the systems of values

$$x = 1, 2, \ldots, p_1, \quad y = 1, 2, \ldots, q_1,$$

independently.

We can never have $qx = py$, because it would follow from this equation that py is a multiple of q which is impossible because $(p, q) = (y, q) = 1$ (since $0 < y < q$). Therefore we can set $p_1 q_1 = S_1 + S_2$, where S_1 is the number of pairs with $qx < py$ and S_2 is the number of pairs with $py < qx$.

It is evident that S_1 is also the number of pairs with $x < \dfrac{p}{q}y$. For given y we can take $x = 1, 2, \ldots, \left[\dfrac{p}{q}y\right]$.

(Since $\dfrac{p}{q}y \leqslant \dfrac{p}{q}q_1 < \dfrac{p}{2}$ we have $\left[\dfrac{p}{q}y\right] \leqslant p_1$.).

Consequently,

$$S_1 = \sum_{y=1}^{q_1} \left[\frac{p}{q}y\right].$$

Analogously, we can prove that

$$S_2 = \sum_{x=1}^{p_1} \left[\frac{q}{p}x\right].$$

But then equation (4) gives

$$\left(\frac{p}{q}\right) = (-1)^{S_1}, \quad \left(\frac{q}{p}\right) = (-1)^{S_2}$$

and hence

$$\left(\frac{p}{q}\right)\left(\frac{q}{p}\right) = (-1)^{S_1+S_2} = (-1)^{p_1 q_1}$$

from which the required property follows.

§3. *The Jacobi Symbol*

a. In order to evaluate Legendre's symbol most quickly, we consider the more general *Jacobi symbol*. Let P be an odd number greater than unity, and let $P = p_1 p_2 \ldots p_r$ be its decomposition into prime factors (some of which may be equal). Moreover, let $(a, P) = 1$. Then Jacobi's symbol is defined by the equation

$$\left(\frac{a}{P}\right) = \left(\frac{a}{p_1}\right)\left(\frac{a}{p_2}\right) \cdots \left(\frac{a}{p_r}\right)$$

The well-known properties of the Legendre symbol allow us to establish analogous properties for the Jacobi symbol.

b. *If* $a \equiv a_1 \pmod{P}$, *then* $\left(\dfrac{a}{P}\right) = \left(\dfrac{a_1}{P}\right)$.

Indeed,

$$\left(\frac{a}{P}\right) = \left(\frac{a}{p_1}\right)\left(\frac{a}{p_2}\right) \cdots \left(\frac{a}{p_r}\right) =$$

$$= \left(\frac{a_1}{p_1}\right)\left(\frac{a_1}{p_2}\right) \cdots \left(\frac{a_1}{p_r}\right) = \left(\frac{a_1}{P}\right)$$

87

so that a, being congruent to a_1 modulo P, is also congruent to a_1 modulo $p_1, p_2, \ldots, p_r$, which are the divisors of P.

c. $\left(\dfrac{1}{P}\right) = 1.$

Indeed,

$$\left(\frac{1}{P}\right) = \left(\frac{1}{p_1}\right)\left(\frac{1}{p_2}\right)\cdots\left(\frac{1}{p_r}\right) = 1.$$

d. $\left(\dfrac{-1}{P}\right) = (-1)^{\frac{P-1}{2}}.$

In order to establish this, we note that

$$(1)\quad \left(\frac{-1}{P}\right) = \left(\frac{-1}{p_1}\right)\left(\frac{-1}{p_2}\right)\cdots\left(\frac{-1}{p_r}\right) =$$

$$= (-1)^{\frac{p_1-1}{2} + \frac{p_2-1}{2} + \cdots + \frac{p_r-1}{2}};$$

but

$$\frac{P-1}{2} = \frac{p_1 p_2 \cdots p_r - 1}{2} =$$

$$= \frac{\left(1 + 2\dfrac{p_1-1}{2}\right)\left(1 + 2\dfrac{p_2-1}{2}\right)\cdots\left(1 + 2\dfrac{p_r-1}{2}\right) - 1}{2} =$$

$$= \frac{p_1-1}{2} + \frac{p_2-1}{2} + \cdots + \frac{p_r-1}{2} + 2N$$

and hence from formula (1) we deduce

$$\left(\frac{-1}{P}\right) = (-1)^{\frac{P-1}{2}}.$$

e. $\left(\dfrac{ab\ldots l}{P}\right) = \left(\dfrac{a}{P}\right)\left(\dfrac{b}{P}\right)\cdots\left(\dfrac{l}{P}\right).$

Indeed,

$$\left(\frac{ab\ldots l}{P}\right) = \left(\frac{ab\ldots l}{p_1}\right)\cdots\left(\frac{ab\ldots l}{p_r}\right) =$$

$$= \left(\frac{a}{p_1}\right)\left(\frac{b}{p_1}\right)\cdots\left(\frac{l}{p_1}\right)\cdots\left(\frac{a}{p_r}\right)\left(\frac{b}{p_r}\right)\cdots\left(\frac{l}{p_r}\right);$$

and multiplying the symbols with the same numerators, we obtain the required property. From this we obtain the corollary:

$$\left(\frac{ab^2}{P}\right) = \left(\frac{a}{P}\right).$$

f. $\left(\dfrac{2}{P}\right) = (-1)^{\frac{P^2-1}{8}}.$

Indeed,

$$\left(\frac{2}{P}\right) = \left(\frac{2}{p_1}\right)\left(\frac{2}{p_2}\right)\cdots\left(\frac{2}{p_r}\right) =$$

(2)

$$= (-1)^{\frac{p_1^2-1}{8} + \frac{p_2^2-1}{8} + \ldots + \frac{p_r^2-1}{8}}.$$

But

$$\frac{P^2-1}{8} = \frac{p_1^2 p_2^2 \ldots p_r^2 - 1}{8} =$$

$$= \frac{\left(1 + 8\dfrac{p_1^2-1}{8}\right)\left(1 + 8\dfrac{p_2^2-1}{8}\right)\cdots\left(1 + 8\dfrac{p_r^2-1}{8}\right) - 1}{8} =$$

$$= \frac{p_1^2-1}{8} + \frac{p_2^2-1}{8} + \ldots + \frac{p_r^2-1}{8} + 2N$$

and hence we deduce from formula (2)

$$\left(\frac{2}{P}\right) = (-1)^{\frac{P^2-1}{8}}.$$

g. *If P and Q are positive relatively prime odd numbers, then*

$$\left(\frac{Q}{P}\right) = (-1)^{\frac{P-1}{2} \cdot \frac{Q-1}{2}} \left(\frac{P}{Q}\right).$$

Indeed, let $Q = q_1 q_2 \ldots q_s$ be the decomposition of Q into prime factors (some of them may be equal). We have

$$\left(\frac{Q}{P}\right) = \left(\frac{Q}{P_1}\right)\left(\frac{Q}{P_2}\right) \ldots \left(\frac{Q}{P_r}\right) = \prod_{\alpha=1}^{r} \prod_{\beta=1}^{s} \left(\frac{q_\beta}{P_\alpha}\right) =$$

$$= (-1)^{\sum\limits_{\alpha=1}^{r} \sum\limits_{\beta=1}^{s} \frac{P_\alpha-1}{2} \cdot \frac{q_\beta-1}{2}} \prod_{\alpha=1}^{r} \prod_{\beta=1}^{s} \left(\frac{P_\alpha}{q_\beta}\right) =$$

$$= (-1)^{\left(\sum\limits_{\alpha=1}^{r} \frac{P_\alpha-1}{2}\right)\left(\sum\limits_{\beta=1}^{s} \frac{q_\beta-1}{2}\right)} \left(\frac{P}{Q}\right).$$

But, as in **d**, we find

$$\frac{P-1}{2} = \sum_{\alpha=1}^{r} \frac{P_\alpha-1}{2} + 2N, \quad \frac{Q-1}{2} = \sum_{\beta=1}^{s} \frac{q_\beta-1}{2} + 2N_1,$$

and hence

$$\left(\frac{Q}{P}\right) = (-1)^{\frac{P-1}{2} \cdot \frac{Q-1}{2}} \left(\frac{P}{Q}\right).$$

90

Example. As an example of the calculation of the Legendre symbol (we will consider it to be a particular case of the Jacobi symbol) we investigate the solutions of the congruence

$$x^2 \equiv 219 \,(\text{mod } 383).$$

We have (applying in sequence the properties **g, b,** the corollary of **e, g, b, e, f, g, b, d**):

$$\left(\frac{219}{383}\right) = -\left(\frac{383}{219}\right) = -\left(\frac{164}{219}\right) = -\left(\frac{41}{219}\right) =$$

$$= -\left(\frac{219}{41}\right) = -\left(\frac{14}{41}\right) = -\left(\frac{2}{41}\right)\left(\frac{7}{41}\right) =$$

$$= -\left(\frac{7}{41}\right) = -\left(\frac{41}{7}\right) = -\left(\frac{-1}{7}\right) = 1,$$

and hence the congruence under consideration has two solutions.

§4. *The Case of Composite Moduli*

a. Congruences of the second degree with composite moduli are investigated and solved in accordance with the general methods of §**5, ch. IV.**

b. We start with a congruence of the form

$$(1) \qquad x^2 \equiv a \,(\text{mod } p^{\alpha}); \ \alpha > 0, \ (a, p) = 1,$$

where p is an odd prime.

Setting $f(x) = x^2 - a$, we have $f'(x) = 2x$, and if $x \equiv x_1$ (mod p) is a solution of the congruence

$$(2) \qquad\qquad x^2 \equiv a \,(\text{mod } p)$$

then since $(a, p) = 1$ we also have $(x_1, p) = 1$, and since p is odd, $(2x_1, p) = 1$, i.e. $f'(x_1)$ is not divisible by p. Therefore to find the solutions of the congruence (1) we can apply the argument of **b, §5, ch. IV**, while each solution of the congruence (2) gives one solution of the congruence (1). It follows from this that

The congruence (1) *has two solutions or none according as a is a quadratic residue or a quadratic non-residue modulo* p.

c. We now consider the congruence

(3) $$x^2 \equiv a \,(\text{mod}\, 2^{\alpha}); \quad \alpha > 0, \quad (a, 2) = 1.$$

Here $f'(x_1) = 2x_1$ is divisible by 2, and hence the argument of **b, §5, ch. IV** is inapplicable; it can be changed in the following way:

d. If the congruence (3) is solvable, then, since $(a, 2) = 1$, we have $(x, 2) = 1$, i.e. $x = 1 + 2t$, where t is an integer. The congruence (2) takes on the form

$$1 + 4t(t + 1) \equiv a \,(\text{mod}\, 2^{\alpha}).$$

But one of the numbers t, $t + 1$ is even and hence $4t(t + 1)$ is a multiple of 8. Therefore, for the solvability of the latter congruence, and along with it also the congruence (3), it is necessary that

(4) $a \equiv 1 \,(\text{mod}\, 4)$ for $\alpha = 2$; $a \equiv 1 \,(\text{mod}\, 8)$ for $\alpha \geqslant 3$.

e. In the cases in which condition (4) is satisfied, we consider the question of finding solutions and the number of solutions.

For $\alpha \leqslant 3$, all the odd numbers satisfy the congruence in view of **d.** Therefore the congruence $x^2 \equiv a \,(\text{mod}\, 2)$ has one solution: $x \equiv 1 \,(\text{mod}\, 2)$, the congruence $x^2 \equiv a \,(\text{mod}\, 4)$ has two solutions: $x \equiv 1; 3 \,(\text{mod}\, 4)$, the congruence $x^2 \equiv a \,(\text{mod}\, 8)$ has four solutions: $x \equiv 1, 3, 5, 7 \,(\text{mod}\, 8)$.

In order to consider the cases $\alpha = 4, 5, \ldots$ all the odd numbers are put in the two arithmetic progressions:

$$(5) \qquad\qquad x = \pm(1 + 4t_3)$$

$$(1 + 4t_3 \equiv 1 \,(\text{mod } 4); \; -1 - 4t_3 \equiv -1 \equiv 3 \,(\text{mod } 4))$$

We now decide which of the latter numbers satisfy the congruence $x^2 \equiv a \,(\text{mod } 16)$. We find

$$(1 + 4t_3)^2 \equiv a \,(\text{mod } 16), \; t_3 \equiv \frac{a - 1}{8} \,(\text{mod } 2),$$

$$t_3 = t_3' + 2t_4, \; x = \pm(1 + 4t_3' + 8t_4) = \pm(x_4 + 8t_4).$$

We now decide which of the latter numbers satisfy the congruence $x^2 \equiv a \,(\text{mod } 32)$. We find

$$(x_4 + 8t_4)^2 \equiv a \,(\text{mod } 32), \; t_4 = t_4' + 2t_5, \; x = \pm(x_5 + 16t_5),$$

etc. In this way we find that the values of x satisfying the congruence (3) for $\alpha > 3$, are representable in the form

$$x = \pm(x_\alpha + 2^{\alpha-1} t_\alpha).$$

These values of x form four different solutions of the congruence (3)

$$x \equiv x_\alpha; \; x_\alpha + 2^{\alpha-1}; \; -x_\alpha; \; -x_\alpha - 2^{\alpha-1} \,(\text{mod } 2^\alpha)$$

(modulo 4 the first two are congruent to 1 while the second two are congruent to -1).

   Example. The congruence

$$(6) \qquad\qquad x^2 \equiv 57 \,(\text{mod } 64)$$

has four solutions since $57 \equiv 1 \,(\text{mod } 8)$. Representing x in the form $x = \pm(1 + 4t_3)$, we find

$$(1 + 4t_3)^2 \equiv 57\,(\text{mod } 16), \quad 8t_3 \equiv 56\,(\text{mod } 16),$$

$$t_3 \equiv 1\,(\text{mod } 2), \quad t_3 = 1 + 2t_4, \quad x = \pm(5 + 8t_4),$$

$$(5 + 8t_4)^2 \equiv 57\,(\text{mod } 32), \quad 5 \cdot 16t_4 \equiv 32\,(\text{mod } 32),$$

$$t_4 \equiv 0\,(\text{mod } 2), \quad t_4 = 2t_5, \quad x = \pm(5 + 16t_5),$$

$$(5 + 16t_5)^2 \equiv 57\,(\text{mod } 64), \quad 5 \cdot 32t_5 \equiv 32\,(\text{mod } 64),$$

$$t_5 \equiv 1\,(\text{mod } 2), \quad t_5 = 1 + 2t_6, \quad x = \pm(21 + 32t_6).$$

Therefore the solutions of the congruence (6) are:

$$x \equiv \pm 21; \quad \pm 53\,(\text{mod } 64).$$

f. It follows from **c, d,** and **e** that:
The necessary conditions for the solvability of the congruence

$$x^2 \equiv a\,(\text{mod } 2^{\alpha}); \quad (a,\ 2) = 1$$

are: $a \equiv 1\,(\text{mod } 4)$ *for* $\alpha = 2$, $a \equiv 1\,(\text{mod } 8)$ *for* $\alpha \geqslant 3$. *If these conditions are satisfied, then the number of solutions is:* 1 *for* $\alpha = 1$; 2 *for* $\alpha = 2$; 4 *for* $\alpha \geqslant 3$.

g. It follows from **b, f** and **a, § 5, ch. IV** that:
Necessary conditions for the solvability of congruences of the form

$$x^2 \equiv a\,(\text{mod } m); \quad m = 2^{\alpha} p_1^{\alpha_1} p_2^{\alpha_2} \dots p_k^{\alpha_k}; \quad (a,\ m) = 1$$

are:

$$a \equiv 1\,(\text{mod } 4) \text{ for } \alpha = 2, \quad a \equiv 1\,(\text{mod } 8) \text{ for } \alpha \geqslant 3,$$

$$\left(\frac{a}{p_1}\right) = 1, \quad \left(\frac{a}{p_2}\right) = 1, \quad \dots, \quad \left(\frac{a}{p_k}\right) = 1.$$

If all of these conditions are satisfied, the number of solutions is: 2^k *for* $\alpha = 0$ *and* $\alpha = 1$; 2^{k+1} *for* $\alpha = 2$; 2^{k+2} *for* $\alpha \geqslant 3$.

94

Problems for Chapter V

Here p will always denote an odd prime.

1. Prove that finding the solutions of a congruence of the form

$$ax^2 + bx + c \equiv 0 \,(\text{mod } m), \quad (2a, \, m) = 1$$

reduces to finding the solutions of a congruence of the form $x^2 \equiv q \,(\text{mod } m)$.

2, a. Using **e, §1,** find the solutions of the congruence (when they exist)

$$x^2 \equiv a \,(\text{mod } p); \quad p = 4m + 3.$$

b. Using **b** and **k, §2,** obtain a method of finding the solutions of the congruence

$$x^2 \equiv a \,(\text{mod } p); \quad p = 8m + 5.$$

c. Find the simplest possible method of finding the solutions of a congruence of the form

$$x^2 \equiv a \,(\text{mod } p); \quad p = 8m + 1$$

when we know some quadratic non-residue N modulo p.

d. Using Wilson's theorem, prove that the solutions of the congruence

$$x^2 + 1 \equiv 0 \,(\text{mod } p); \quad p = 4m + 1$$

are

$$x \equiv \pm 1 \cdot 2 \ldots 2m \,(\text{mod } p).$$

3, a. Prove that the congruence

(1) $$x^2 + 1 \equiv 0 \,(\text{mod } p)$$

is solvable if and only if p is of the form $4m + 1$; the congruence

(2) $$x^2 + 2 \equiv 0 \,(\text{mod } p)$$

is solvable if and only if p is of the form $8m + 1$ or $8m + 3$; the congruence

$$x^2 + 3 \equiv 0 \,(\text{mod } p)$$

is solvable if and only if p is of the form $6m + 1$.

b. Prove that there are an infinite number of primes of the form $4m + 1$.

c. Prove that there are an infinite number of primes of the form $6m + 1$.

4. Dividing the numbers $1, 2, \ldots, p - 1$ into two sets, the second of which contains at least one number, we assume that the product of two numbers of the same set are congruent to a number of the first set modulo p, while the product of two elements of different sets is congruent to a number of the second set modulo p. Prove that this occurs if and only if the first set consists of quadratic residues, while the second set consists of quadratic non-residues modulo p.

5, a. Deduce the theory of congruences of the form

$$x^2 \equiv a \,(\text{mod } p^{\alpha}); \; (a, \, p) = 1,$$

by representing a and x in the calculational system to the base p.

b. Deduce the theory of congruences of the form

$$x^2 \equiv a \,(\text{mod } 2^{\alpha}); \; (a, \, 2) = 1,$$

by representing a and x in the calculational system to the base 2.

6. Prove that the solutions of the congruence

$$x^2 \equiv a \,(\text{mod } p^{\alpha}); \; (a, \, p) = 1$$

are $x \equiv \pm PQ' \pmod{p^{\alpha}}$, where

$$P = \frac{(z + \sqrt{a})^{\alpha} + (z - \sqrt{a})^{\alpha}}{2}, \quad Q = \frac{(z + \sqrt{a})^{\alpha} - (z - \sqrt{a})^{\alpha}}{2\sqrt{a}}$$

$$z^2 \equiv a \pmod{p}, \quad QQ' \equiv 1 \pmod{p^{\alpha}}.$$

7. Find a method of solving the congruence $x^2 \equiv 1 \pmod{m}$ based on the fact that this congruence is equivalent to the congruence $(x - 1)(x + 1) \equiv 0 \pmod{m}$.

8. Let $\left(\dfrac{a}{p}\right) = 0$ for $(a, p) = p$.

a. For $(k, p) = 1$, prove that

$$\sum_{x=0}^{p-1} \left(\frac{x(x + k)}{p}\right) = -1.$$

b. Let each of the numbers ϵ and η have one of the values ± 1, let T be the number of pairs $x, x + 1$, where $x = 1, 2,$ $\ldots, p - 2$, such that $\left(\dfrac{x}{p}\right) = \epsilon$, $\left(\dfrac{x + 1}{p}\right) = \eta$. Prove that

$$T = \left(\frac{1}{4} \left(p - 2 - \epsilon\left(\frac{-1}{p}\right) - \eta - \epsilon\eta\right)\right)$$

c. Let $(k, p) = 1$, and let

$$S = \sum_{x}\sum_{y} \left(\frac{xy + k}{p}\right)$$

where x and y run through increasing sequences consisting, respectively, of X and Y residues of a complete system modulo

p. Prove that

$$|S| < \sqrt{2XYp}$$

In the proof use the inequality

$$|S|^2 \leqslant X \sum_x \left| \sum_y \left(\frac{xy + k}{p} \right) \right|^2$$

d. Let Q be an integer, $1 < Q < p$,

$$S = \sum_{x=0}^{p-1} S_x^2; \quad S_x = \sum_{z=0}^{Q-1} \left(\frac{x + z}{p} \right).$$

α) Prove that $S = (p - Q)Q$.

β) Let λ be a constant, $0 < \lambda < 1$. Prove that the number T of integers $x = 0, 1, \ldots, p - 1$ for which the condition $S_x < Q^{0.5 + 0.5\lambda}$ is not satisfied, satisfies the condition $T \leqslant pQ^{-\lambda}$.

γ) Let $p > 25$, and let M be an integer. Prove that the sequence

$$M, M + 1, \ldots, M + 3[\sqrt{p}] - 1$$

contains a quadratic non-residue modulo p.

9, a. Prove that the number of representations of an integer $m > 1$ in the form

(1) $m = x^2 + y^2, \ (x, y) = 1, \ x > 0, \ y > 0$

is equal to the number of solutions of the congruence

(2) $z^2 + 1 \equiv 0 \,(\text{mod } m).$

In proving this, set $r = \sqrt{m}$ and use the representation of $\alpha = \dfrac{z}{m}$ given in the theorem of problem **4, b, ch. I,** and then consider the congruence obtained by multiplying (2) termwise by Q^2.

b. Let a be one of the numbers 2, 3. Prove that the number of representations of a prime $p > a$ in the form

$$(3) \qquad\qquad p = x^2 + ay^2, \; x > 0, \; y > 0$$

is equal to half the number of solutions of the congruence

$$(4) \qquad\qquad z^2 + a \equiv 0 (\mathrm{mod}\; p).$$

c. Let p be of the form $4m + 1$, $(k, p) = 1$,

$$S(k) = \sum_{x=0}^{p-1} \left(\frac{x(x^2 + k)}{p} \right)$$

Prove that (D. S. Gorshkov)

$\alpha)$ $S(k)$ is an even number.

$\beta)$ $S(kt) = \left(\dfrac{t}{p} \right) S(k).$

$\gamma)$ For $\left(\dfrac{r}{p} \right) = 1$, $\left(\dfrac{n}{p} \right) = -1$, we have (cf. problem **a.**)

$$p = \left(\frac{1}{2} S(r) \right)^2 + \left(\frac{1}{2} S(n) \right)^2.$$

10. Let D be a positive integer which is not the square of an integer. Prove that:

a. If two pairs $x = x_1, \; y = y_1$ and $x = x_2, \; y = y_2$ of integers satisfy the equation

$$x^2 - Dy^2 = k$$

99

for a given integer k, then the equation

$$X^2 - DY^2 = k^2$$

is satisfied by integers X, Y defined by the equation (the $\pm$ sign can be taken arbitrarily)

$$X + Y\sqrt{D} = (x_1 + y_1\sqrt{D})(x_2 \pm y_2\sqrt{D}).$$

b. The equation (Pell's equation)

(1) $$x^2 - Dy^2 = 1$$

is solvable in positive integers x, y.

c. If x_0, y_0 is a pair of positive x, y with minimal x (or, equivalently, with minimal $x + y\sqrt{D}$) satisfying equation (1), then all pairs of positive x, y satisfying this equation are defined by the equations

(2) $$x + y\sqrt{D} = (x_0 + y_0\sqrt{D})^r; \quad r = 1, 2, \ldots$$

11, a. Let a be an integer. Let

$$U_{a,p} = \sum_{x=1}^{p-1} \left(\frac{x}{p}\right) e^{2\pi i \frac{ax}{p}}.$$

α) For $(a, p) = 1$, prove that $\left| U_{a,p} \right| = \sqrt{p}$.

In proving this, multiply $U_{a,p}$ by its conjugate, which is obtained by replacing i by $-i$. Letting the letters x_1 and x be the summation variables of the original and conjugate sums, we then gather together the terms of the product such that

$$x_1 \equiv xt \,(\text{mod } p),$$

or

$$x_1 \equiv x + t \,(\text{mod } p)$$

100

for fixed t.

β) Prove that

$$\left(\frac{a}{p} \right) = \frac{U_{a,p}}{U_{1,p}}$$

b. Let $m > 2$, $(a, m) = 1$,

$$S_{a,m} = \sum_{x=0}^{m-1} e^{2\pi i \frac{a x^2}{m}}$$

α) Prove that $S_{a,p} = U_{a,p}$ (problem **a**).

β) It follows from the theorems of problems α) and **a**, α) that $S_{a,p} = \sqrt{p}^{*}$. Prove the following more general result:

$$|S_{a,m}| = \sqrt{m}, \quad \text{if } m \equiv 1 \,(\text{mod } 2),$$
$$|S_{a,m}| = 0, \qquad \text{if } m \equiv 2 \,(\text{mod } 4),$$
$$|S_{a,m}| = \sqrt{2m}, \quad \text{if } m \equiv 0 \,(\text{mod } 4).$$

γ) Let $m > 1$, $(2A, m) = 1$, and let a be an arbitrary integer. Prove that

$$\left| \sum_{x=0}^{m-1} \exp\left(2\pi i \frac{A x^2 + ax}{m} \right) \right| = \sqrt{m}.$$

12, a. Let m be an integer exceeding 1, let M and Q be integers such that $0 \leqslant M < M + Q \leqslant m$, and let $\sum_{z}$ denote a sum extended over the z in a given set of integers, while $\sum_{z}'$ denotes a sum extended over the z in this set which are congruent modulo m to the numbers

$$M, M + 1, \ldots, M + Q - 1.$$

Moreover, let the function $\Phi(z)$ be such that, for some Δ and any $a = 1, 2, \ldots, m - 1$, we have

$$\left| \sum_z \Phi(z) \exp\left(2\pi i \frac{az}{m}\right) \right| \leqslant \Delta.$$

Prove that

$$\sum_z{}' \Phi(z) = \frac{Q}{m} \sum_z \Phi(z) + \theta\Delta(\ln m - \delta),$$

where $|\theta| < 1$, $\delta > 0$ always, $\delta > \dfrac{1}{2}$ for $m \geqslant 12$,

$\delta > 1$ for $m \geqslant 60$.

b. Let M and Q be integers such that $0 < M < M + Q \leqslant p$.
α) Prove that

$$\left| \sum_{x=M}^{M+Q-1} \left(\frac{x}{p}\right) \right| < \sqrt{p}\, \ln p.$$

β) Let R be the number of quadratic residues and let N be the number of quadratic non-residues in the sequence M, $M + 1, \ldots, M + Q - 1$. Prove that

$$R = \frac{1}{2}Q + \frac{\theta}{2}\sqrt{p}\, \ln p, \quad N = \frac{1}{2}Q - \frac{\theta}{2}\sqrt{p}\, \ln p; \quad |\theta| < 1.$$

γ) Deduce the formulae of problem β), using the theorem of problem **11, b,** β) and the theorem of problem **a.**
δ) Let $m > 2$, $(2A, m) = 1$, and let M_0 and Q_0 be integers such that $0 < M_0 < M_0 + Q_0 \leqslant m$. Prove that

$$\left| \sum_{x=M_0}^{M_0+Q_0-1} \exp\left(2\pi i \frac{Ax^2}{m}\right) \right| < \sqrt{m} \ln m.$$

ϵ) Let $p > 2$, $(A, p) = 1$, let M_0 and Q_0 be integers such that $0 < M_0 < M_0 + Q_0 \leqslant p$ and let T be the number of integers of the sequence Ax^2; $x = M_0, M_0 + 1, \ldots, M_0 + Q_0 - 1$, which are congruent modulo p to the numbers of the sequence $M, M + 1, \ldots, M + Q - 1$. Prove that

$$T = \frac{Q_0 Q}{p} + \theta \sqrt{p} \ (\ln p)^2.$$

c. Deduce the formulae of problem **b**, β) by considering the sum

$$\sum_{a=0}^{p-1} \sum_{\alpha=1}^{p-1} \sum_{x=M}^{M+Q-1} \sum_{y=M}^{M+Q-1} \left(\frac{\alpha}{p}\right) \exp\left(2\pi i \frac{a(x - \alpha y)}{p}\right)$$

Numerical Exercises for Chapter V

1, a. Find the quadratic residues in a reduced residue system modulo 23.

b. Find the quadratic non-residues in a reduced residue system modulo 37.

2, a. Applying **e**, §1, find the number of solutions of the congruences:

α) $x^2 \equiv 3 \, (\text{mod} \, 31)$; β) $x^2 \equiv 2 \, (\text{mod} \, 31)$.

b. Find the number of solutions of the congruences:

α) $x^2 \equiv 5 \, (\text{mod} \, 73)$; β) $x^2 \equiv 3 \, (\text{mod} \, 73)$.

3, a. Using the Jacobi symbol, find the number of solutions of the congruences:

α) $x^2 \equiv 226 \,(\text{mod } 563)$; β) $x^2 \equiv 429 \,(\text{mod } 563)$.

b. Find the number of solutions of the congruences:

α) $x^2 \equiv 3766 \,(\text{mod } 5987)$; β) $x^2 \equiv 3149 \,(\text{mod } 5987)$.

4, a. Applying the methods of problems **2, a; 2, b; 2, c,** solve the congruences:

α) $x^2 \equiv 5 \,(\text{mod } 19)$; β) $x^2 \equiv 5 \,(\text{mod } 29)$; γ) $x^2 \equiv 2 \,(\text{mod } 97)$.

b. Solve the congruences:

α) $x^2 \equiv 2 \,(\text{mod } 311)$; β) $x^2 \equiv 3 \,(\text{mod } 277)$;

γ) $x^2 \equiv 11 \,(\text{mod } 353)$.

5, a. Solve the congruence $x^2 \equiv 59 \,(\text{mod } 125)$ by the methods of:

α) **b, $4**; β) problem **5, a**; γ) problem **6**.

b. Solve the congruence $x^2 \equiv 91 \,(\text{mod } 243)$.

6, a. Solve the congruence $x^2 \equiv 41 \,(\text{mod } 64)$ by the methods of:

α) **e, $4**; β) problem **5, b**.

b. Solve the congruence $x^2 \equiv 145 \,(\text{mod } 256)$.

CHAPTER VI
PRIMITIVE ROOTS AND INDICES

§1. General Theorems

a. For $(a, m) = 1$ there exist positive γ such that $a^\gamma \equiv 1$ (mod m), for example (by Euler's theorem) $\gamma = \varphi(m)$. The smallest of these is called: *the exponent to which a belongs modulo m.*

b. *If a belongs to the exponent δ modulo m, then the numbers* $1 = a^0, a^1, \ldots, a^{\delta-1}$ *are incongruent modulo m.*

Indeed, it would follow from $a^l \equiv a^k$ (mod m), $0 \leqslant k < l < \delta$ that $a^{l-k} \equiv 1$ (mod m), $0 < l - k < \delta$, which contradicts the definition of δ.

c. *If a belongs to the exponent δ modulo m, then $a^\gamma \equiv a^{\gamma'}$* (mod m) *if and only if $\gamma \equiv \gamma'$ (mod δ); in particular (for $\gamma' = 0$), $a^\gamma \equiv 1$ (mod m) if and only if γ is divisible by δ.*

Indeed, let r and r_1 be the least non-negative residues of the numbers γ and γ' modulo δ; then for some q and q_1 we have $\gamma = \delta q + r$, $\gamma' = \delta q_1 + r$. From this and from $a^\delta \equiv 1$ (mod m) it follows that

$$a^\gamma = (a^\delta)^q a^r \equiv a^r \,(\text{mod } m),$$

$$a^{\gamma'} = (a^\delta)^{q_1} a^{r_1} \equiv a^{r_1} \,(\text{mod } m).$$

Therefore $a^\gamma \equiv a^{\gamma_1}$ (mod m) if and only if $a^r \equiv a^{r_1}$ (mod m), i.e. (**b**), when $r = r_1$.

d. It follows from $a^{\varphi(m)} \equiv 1 \pmod{m}$ and from **c** ($\gamma' = 0$) that $\varphi(m)$ is divisible by δ. Thus *the exponents to which numbers belong modulo m are just the divisors of* $\varphi(m)$. The largest of these divisors is $\varphi(m)$. The numbers belonging to the exponent $\varphi(m)$ (if such exist) are called the *primitive roots modulo m*.

§2. *Primitive Roots Modulo* p^α *and* $2p^\alpha$

a. Let p be an odd prime and let $\alpha \geqslant 1$. We shall prove the existence of primitive roots modulo p^α and $2p^\alpha$.

b. *If x belongs to the exponent ab modulo m, then x^a belongs to the exponent b.*

Indeed, let x^a belong to the exponent δ. Then $x^{a\delta} \equiv 1$ (mod m), and hence (**c, §1**) $a\delta$ is divisible by ab, i.e. δ is divisible by b. On the other hand, $(x^a)^b \equiv 1 \pmod{m}$ implies (**c, §1**) that b is divisible by δ. Hence $\delta = b$.

c. *If x belongs to the exponent a, and y belongs to the exponent b modulo m, where $(a, b) = 1$, then xy belongs to the exponent ab.*

Indeed, let xy belong to the exponent δ. Then $(xy)^\delta \equiv 1$ (mod m). Hence $x^{b\delta} y^{b\delta} \equiv 1 \pmod{m}$ and (**c, §1**) $x^{b\delta} \equiv 1$ (mod m). Hence (**c, §1**) $b\delta$ is divisible by a, and since $(b, a) = 1$, δ is divisible by a. In the same way we find that δ is divisible by b. Since $(a, b) = 1$, being divisible by a and b, δ is also divisible by ab. On the other hand, $(xy)^{ab} \equiv 1 \pmod{m}$ implies (**c, §1**) that ab is divisible by δ. Hence $\delta = ab$.

d. *There exist primitive roots modulo p.*

Indeed, let τ be the least common multiple of all those exponents

$$(1) \qquad\qquad \delta_1, \delta_2, \ldots, \delta_r,$$

to each of which belongs at least one number of the sequence $1, 2, \ldots, p - 1$ modulo p, and let $\tau = q_1^{\alpha_1} q_2^{\alpha_2} \ldots q_k^{\alpha_k}$ be the canonical decomposition of the number τ. Then for each s,

106

among the numbers (1) there exists some δ which is divisible by $q_s^{\alpha_s}$ and is therefore representable in the form $\delta = a q_s^{\alpha_s}$. If x is a number belonging to the exponent δ, then, by **b**, $x_s = x^a$ belongs to the exponent $q_s^{\alpha_s}$. This holds for $s = 1, 2, \ldots, k$; by **c**, the number $g = x_1 x_2 \ldots x_k$ belongs to the exponent $q_1^{\alpha_1} q_2^{\alpha_2} \ldots q_k^{\alpha_k} = r$.

But since the exponents (1) are just the divisors of the number r, all the numbers $1, 2, \ldots, p - 1$ satisfy (**c**, §1) the congruence $x^r \equiv 1 \pmod p$. This means (**c**, §4, ch. **IV**) that $p - 1 \leqslant r$. But r is a divisor of $p - 1$. Hence $r = p - 1$, i.e. g is a primitive root.

e. *Let g be a primitive root modulo p. We can find a t such that u, which is defined by the equation $(g + pt)^{p-1} = = 1 + pu$, is not divisible by p. The corresponding $g + pt$ is a primitive root modulo p^α for any $\alpha > 1$.*

Indeed, we have

(2)
$$g^{p-1} = 1 + pT,$$

$$(g + pt)^{p-1} = 1 + p(T_0 - g^{p-2}t + pT) = 1 + pu,$$

where, along with t, u runs through a complete residue system modulo p. Therefore, we can find a t such that u is not divisible by p. For this t, we deduce from (2) the equations

(3)
$$
\begin{cases}
(g + pt)^{p\,(p-1)} = (1 + pu)^p = 1 + p^2 u_2, \\
(g + pt)^{p^2\,(p-1)} = (1 + p^2 u_2)^p = 1 + p^3 u_3, \\
\cdots\cdots\cdots\cdots\cdots\cdots\cdots\cdots\cdots\cdots\cdots\cdots\cdots
\end{cases}
$$

where $u_2, u_3, \ldots$ are not divisible by p.

Let $g + pt$ belong to the exponent δ modulo p^α. Then

(4)
$$(g + pt)^\delta \equiv 1 \pmod{p^\alpha}.$$

Hence $(g + pt)^\delta \equiv 1 \pmod p$; and consequently δ is a multiple of $p - 1$, and since δ divides $\varphi(p^\alpha) = p^{\alpha-1}(p - 1)$,

107

it follows that $\delta = p^{r-1}(p - 1)$, where r is one of the numbers $1, 2, \ldots, \alpha$. Replacing the left side of the congruence (4) by its expression in the appropriate equation of (2) or (3), we find $(u = u_1)$

$$1 + p^r u_r \equiv 1 \,(\mathrm{mod}\, p^\alpha),\ p^r \equiv 0 \,(\mathrm{mod}\, p^\alpha),\ r = \alpha,\ \delta = \varphi(p^\alpha),$$

i.e. $g + pt$ is a primitive root modulo p^α.

f. *Let $\alpha \geqslant 1$ and let g be a primitive root modulo p^α. Whichever of the numbers g and $g + p^\alpha$ is odd, is a primitive root modulo $2p^\alpha$.*

Indeed, every odd x which satisfies one of the congruences $x^\gamma \equiv 1 \,(\mathrm{mod}\, p^\alpha)$ and $x^\gamma \equiv 1 \,(\mathrm{mod}\, 2p^\alpha)$ obviously satisfies the other also. Hence, since $\varphi(p^\alpha) = \varphi(2p^\alpha)$ for all odd x, a primitive root for one of the moduli p^α and $2p^\alpha$, is also a primitive root for the other. But, of the two primitive roots g and $g + p^\alpha$ modulo p^α, at least one is odd; and consequently, it will be a primitive root modulo $2p^\alpha$.

§3. *Evaluation of the Primitive Roots*
for the Moduli p^α and $2p^\alpha$

The primitive roots for the moduli p^α and $2p^\alpha$ where p is an odd prime and $\alpha \geqslant 1$, can be found by using the following general theorem:

Let $c = \varphi(m)$ and let $q_1, q_2, \ldots, q_k$ be the different prime divisors of the number c. In order that a number g, which is relatively prime to m, be a primitive root modulo m, it is necessary and sufficient that this g satisfy none of the congruences

$$g^{\frac{c}{q_1}} \equiv 1 \,(\mathrm{mod}\, m),\ g^{\frac{c}{q_2}} \equiv 1 \,(\mathrm{mod}\, m),$$

(1)

$$\ldots,\ g^{\frac{c}{q_k}} \equiv 1 \,(\mathrm{mod}\, m).$$

108

Indeed, if g is a primitive root, then a fortiori it belongs to the exponent c and hence none of the congruences of (1) can be satisfied.

Conversely, we now assume that g satisfies none of the congruences of (1). If the exponent δ to which g belongs, turns out to be less than c, then, letting q be one of the prime divisors of $\dfrac{c}{\delta}$, we would have $\dfrac{c}{\delta} = qu$, $\dfrac{c}{q} = \delta u$, $g^{\frac{c}{q}} \equiv 1$ (mod p), which contradicts our assumption. Hence $\delta = c$ and g is a primitive root.

Example 1. Let $m = 41$. We have $\varphi(41) = 40 = 2^3 \cdot 5$, $\dfrac{40}{5} = 8$, $\dfrac{40}{2} = 20$. Therefore, in order that the number g, not divisible by 41, be a primitive root modulo 41, it is necessary and sufficient that this g satisfy neither of the congruences

(2) $g^8 \equiv 1 \,(\text{mod } 41)$, $g^{20} \equiv 1 \,(\text{mod } 41)$.

But going through the numbers 2, 3, 4, ... we find (modulo 41)

$$2^8 \equiv 10, \ 3^8 \equiv 1, \ 4^8 \equiv 18, \ 5^8 \equiv 18, \ 6^8 \equiv 10,$$

$$2^{20} \equiv 1, \qquad\quad 4^{20} \equiv 1, \ \ 5^{20} \equiv 1, \ \ 6^{20} \equiv 40.$$

From this we see that the numbers 2, 3, 4, 5 are not primitive roots since each of them satisfies at least one of the congruences (2). The number 6 is a primitive root since it satisfies neither of the congruences of (2).

Example 2. Let $m = 1681 = 41^2$. A primitive root can also be obtained here by using the general theorem. But we can find it more simply by applying theorem e, §2. Knowing (example 1) that 6 is a primitive root modulo 41, we find

$$6^{40} = 1 + 41(3 + 41l)$$

$$(6 + 41t)^{40} = 1 + 41(3 + 41l - 6^{39}t + 41T) = 1 + 41u.$$

In order that u be non-divisible by 41, it is sufficient to take $t = 0$. We can therefore take the number $6 + 41 \cdot 0 = 6$ as a primitive root modulo 1681.

Example 3. Let $m = 3362 = 2 \cdot 1681$. The primitive root can also be·obtained here by using the general theorem. But we can find it more simply by applying theorem **f, §2**. Knowing (example 2) that 6 is a primitive root modulo 1681, we can take as a primitive root modulo 3362 the odd number in the pair 6, 6 + 1681, i.e. the number 1687.

§4. *Indices for the Moduli* p^{α} *and* $2p^{\alpha}$

a. Let p be an odd prime, $\alpha \geqslant 1$; let m be one of the numbers p^{α} and $2p^{\alpha}$; $c = \varphi(m)$, and let g be a primitive root modulo m.

b. *If γ runs through the least non-negative residues $\gamma = 0, 1, \ldots, c - 1$ modulo c, then g^{γ} runs through a reduced residue system modulo m.*

Indeed, g^{γ} runs through c numbers which are relatively prime to m, and by **b, §1**, incongruent modulo m.

c. For numbers a, which are relatively prime to m, we introduce the concept of index, which is analogous to the concept of logarithm; here, a primitive root plays a role analogous to the role of the base of a logarithm:

If

$$a \equiv g^{\gamma} \pmod{m}$$

(we assume that $\gamma \geqslant 0$), then γ is said to be the *index of the number a modulo m to the base g* and is denoted by the symbol $\gamma = \operatorname{ind} a$ (more precisely: $\gamma = \operatorname{ind}_g a$).

In view of **b**, every a, relatively prime to m, has some unique index γ' among the numbers of the sequence

$$\gamma = 0, 1, \ldots, c - 1.$$

Knowing γ', we can find all the indices of the number a; by **c, §1**, these are all the non-negative numbers of the class

110

$$\gamma \equiv \gamma'(\bmod c).$$

It follows immediately from the definition of the index which we have given here that the numbers with a given index γ form an equivalence class of numbers modulo m.

d. *We have*

$$\text{ind } ab \ldots l \equiv \text{ind } a + \text{ind } b + \ldots + \text{ind } l \ (\bmod c)$$

and in particular,

$$\text{ind } a^n \equiv n \text{ ind } a \ (\bmod c).$$

Indeed,

$$a \equiv g^{\text{ind } a} \ (\bmod m), \ b \equiv g^{\text{ind } b} \ (\bmod m),$$

$$\ldots, \ l \equiv g^{\text{ind } l} \ (\bmod m),$$

and multiplying the latter together, we find

$$ab \ldots l \equiv g^{\text{ind } a + \text{ind } b + \ldots + \text{ind } l} \ (\bmod m).$$

Therefore, ind a + ind b + $\ldots$ + ind l is one of the indices of the product $ab \ldots l$.

e. In view of the practical use of indices, for each prime modulus p (which is not too large) tables of indices have been constructed. There are two tables: one for finding the index from the number, and the other for finding the number from the index. The tables contain the least non-negative residues of the numbers (a reduced residue system) and their smallest indices (a complete system) corresponding to a modulus p and $c = \varphi(p) = p - 1$.

Example. We construct the preceding table for the modulus $p = 41$. It was shown above (example 1, §3) that $g = 6$ is a primitive root modulo 41; we take it as the basis of the indices. We find (congruences are taken modulo 41):

$$6^0 \equiv 1 \quad 6^8 \equiv 10 \quad 6^{16} \equiv 18 \quad 6^{24} \equiv 16 \quad 6^{32} \equiv 37$$
$$6^1 \equiv 6 \quad 6^9 \equiv 19 \quad 6^{17} \equiv 26 \quad 6^{25} \equiv 14 \quad 6^{33} \equiv 17$$
$$6^2 \equiv 36 \quad 6^{10} \equiv 32 \quad 6^{18} \equiv 33 \quad 6^{26} \equiv 2 \quad 6^{34} \equiv 20$$
$$6^3 \equiv 11 \quad 6^{11} \equiv 28 \quad 6^{19} \equiv 34 \quad 6^{27} \equiv 12 \quad 6^{35} \equiv 38$$
$$6^4 \equiv 25 \quad 6^{12} \equiv 4 \quad 6^{20} \equiv 40 \quad 6^{28} \equiv 31 \quad 6^{36} \equiv 23$$
$$6^5 \equiv 27 \quad 6^{13} \equiv 24 \quad 6^{21} \equiv 35 \quad 6^{29} \equiv 22 \quad 6^{37} \equiv 15$$
$$6^6 \equiv 39 \quad 6^{14} \equiv 21 \quad 6^{22} \equiv 5 \quad 6^{30} \equiv 9 \quad 6^{38} \equiv 8$$
$$6^7 \equiv 29 \quad 6^{15} \equiv 3 \quad 6^{23} \equiv 30 \quad 6^{31} \equiv 13 \quad 6^{39} \equiv 7$$

and hence our tables are:

N	0	1	2	3	4	5	6	7	8	9
0		0	26	15	12	22	1	39	38	30
1	8	3	27	31	25	37	24	33	16	9
2	34	14	29	36	13	4	17	5	11	7
3	23	28	10	18	19	21	2	32	35	6
4	20									

I	0	1	2	3	4	5	6	7	8	9
0	1	6	36	11	25	27	39	29	10	19
1	32	28	4	24	21	3	18	26	33	34
2	40	35	5	30	16	14	2	12	31	22
3	9	13	37	17	20	38	23	15	8	7

Here the row number is the first digit and the column number is the second digit of the number (index). At the place common to the given row and given column we place the corresponding index (number).

For example, we find the *ind 25* at the place in the first table common to the 2-nd row and the 5-th column, i.e. ind 25 = 4. The number whose index is 33 is found in the place in the second table common to the 3-rd row and the 3-rd column, i.e. 33 = ind 17.

112

§5. *Consequences of the Preceding Theory*

a. Let p be an odd prime; $\alpha \geqslant 1$, let m be one of the numbers p^{α}, $2p^{\alpha}$, and finally, let $c = \varphi(m)$.

b. *Let* $(n, c) = d$; *then:*

1. *The congruence*

(1) $$x^n \equiv a \;(\text{mod } m)$$

is solvable (and hence a is an n-th power residue modulo m) if and only if ind a is a multiple of d.

In the case of solvability the congruence has d solutions.

2. *The number of n-th power residues in a reduced residue system modulo m is* $\dfrac{c}{d}$.

Indeed, the congruence (1) is equivalent to the congruence

(2) $$n \text{ ind } x \equiv \text{ind } a \;(\text{mod } c)$$

which is solvable if and only if *ind a* is a multiple of d (**d, §2, ch. IV**).

If the congruence (2) is solvable, we find d values of *ind x* which are incongruent modulo c; corresponding to them we find d values of x which are incongruent modulo m, proving assertion **1.**

Among the numbers $0, 1, \ldots, c - 1$, which are the smallest indices of a reduced residue system modulo m, there are $\dfrac{c}{d}$ which are multiples of d, proving assertion **2.**

Example **1.** For the congruence

(3) $$x^8 \equiv 23 \;(\text{mod } 41)$$

we have $(8, 40) = 8$, while ind $23 = 36$ is not divisible by 8. Therefore the congruence (3) is unsolvable.

Example 2. For the congruence

(4) $$x^{12} \equiv 37 \pmod{41}$$

we have $(12, 40) = 4$, while ind $37 = 32$ is divisible by 4.
Therefore the congruence (4) is solvable and has 4 solutions.
These solutions are obtained in the following way:
The congruence (4) is equivalent to the following ones:

$$12 \text{ ind } x \equiv 32 \pmod{40}, \text{ ind } x \equiv 6 \pmod{10}.$$

Hence we find 4 values of *ind* x which are incongruent
modulo 40:

$$\text{ind } x = 6, \ 16, \ 26, \ 36,$$

from which we obtain the 4 solutions of the congruence (4)

$$x \equiv 39, \ 18, \ 2, \ 23 \pmod{41}.$$

Example 3. The numbers

(5) $$1, \ 4, \ 10, \ 16, \ 18, \ 23, \ 25, \ 31, \ 37, \ 40$$

whose indices are multiples of 4, are just all the biquadratic
residues (or the residues of any power $n = 12, 28, 36, \ldots$,
where $(n, 40) = 4$), among the least positive residues modulo
41. The number of integers in the sequence (5) is $10 = \dfrac{40}{4}$.

c. Along with assertion **b, 1**, we shall also find the following one useful:
*The number a is an n-th power residue modulo m if and
only if*

(6) $$a^{\frac{c}{d}} \equiv 1 \pmod{m}.$$

114

Indeed, the condition ind $a \equiv 0 \pmod{d}$ is equivalent to the condition: $\dfrac{c}{d}$ ind $a \equiv 0 \pmod{c}$. The latter is equivalent to condition (6).

Example. By the theorem of §3, the impossibility of the congruence $g^{\frac{c}{q}} \equiv 1 \pmod{m}$ is equivalent to the statement that g is a q-th power non-residue modulo m. In particular, the impossibility of the congruence $g^{\frac{c}{2}} \equiv 1 \pmod{m}$ is equivalent to the statement that g is a quadratic non-residue modulo m (cf. **e**, §1, ch. **V**).

d, 1. *The exponent δ to which a belongs modulo m is defined by the equation $(\text{ind } a, c) = \dfrac{c}{\delta}$; in particular, the fact that a belongs to a number of primitive roots modulo m is equivalent to the equation* $(\text{ind } a, c) = 1$.

2. *In a reduced residue system modulo m, the number of numbers belonging to the exponent δ is $\varphi(\delta)$; in particular, the number of primitive roots is $\varphi(c)$.*

Indeed, δ is the smallest divisor of c such that $a^{\delta} \equiv 1 \pmod{m}$. This condition is equivalent to

$$\delta \text{ ind } a \equiv 0 \pmod{c},$$

or

$$\text{ind } a \equiv 0 \left(\text{mod } \frac{c}{\delta}\right).$$

This means that δ is the smallest divisor of c for which $\dfrac{c}{\delta}$ divides *ind* a, from which it follows that $\dfrac{c}{\delta}$ is the largest divisor of c which divides *ind* a, i.e. $\dfrac{c}{\delta} = (\text{ind } a, c)$, proving assertion **1.**

115

Among the numbers $0, 1, \ldots, c - 1$, which are the smallest indices of a reduced residue system modulo m, the multiples of $\dfrac{c}{\delta}$ are the numbers of the form $\dfrac{c}{\delta} y$, where $y = 0, 1,$ $\ldots, \delta - 1$. The condition $\left(\dfrac{c}{\delta} y,\right) c = \dfrac{c}{\delta}$ is equivalent to the condition $(y, \delta) = 1$; and the latter is satisfied by $\varphi(\delta)$ values of y, proving assertion 2.

Example 1. In a reduced residue system modulo 41, the numbers belonging to the exponent 10 are the numbers a such that $(\text{ind } a, 40) = \dfrac{40}{10} = 4$, i.e. the numbers

$$4, \ 23, \ 25, \ 31.$$

The number of these numbers is $4 = \varphi(10)$.

Example 2. In a reduced residue system modulo 41, the primitive roots are the numbers a such that $(\text{ind } a, 40) = 1$, i.e. the numbers

$$6, 7, 11, 12, 13, 15, 17, 19, 22, 24, 26, 28, 29, 30, 34, 35.$$

The number of these primitive roots is $16 = \varphi(40)$.

§6. *Indices Modulo* 2^{α}

a. The preceding theory is replaced, for the modulus 2^{α}, by a somewhat more complicated one.

b. Let $\alpha = 1$. Then $2^{\alpha} = 2$. We have $\varphi(2) = 1$. A primitive root modulo 2 is, for example, $1 \equiv -1 \pmod 2$. The number $1^{0} = (-1)^{0} = 1$ forms a reduced residue system modulo 2.

c. Let $\alpha = 2$. Then $2^{2} = 4$. We have $\varphi(4) = 2$. The number $3 \equiv -1 \pmod 4$ is a primitive root modulo 4. The numbers $(-1)^{0} = 1$, $(-1)^{1} \equiv 3 \pmod 4$ form a reduced residue system modulo 4.

116

d. Let $\alpha \geqslant 3$. Then $2^\alpha \geqslant 8$. We have $\varphi(2^\alpha) = 2^{\alpha-1}$. It is easy to see that there are no primitive roots in this case; more precisely: the exponent to which the odd number x belongs modulo 2^α does not exceed $2^{\alpha-2} = \dfrac{1}{2}\,\varphi(2^\alpha)$. Indeed, we have

$$x^2 = 1 + 8t_1,$$

$$x^4 = 1 + 16t_2,$$

$$\dotfill$$

$$x^{2^{\alpha-2}} = 1 + 2^\alpha t_{\alpha-2} \equiv 1 \;(\mathrm{mod}\; 2^\alpha).$$

Therefore, there exist numbers belonging to the exponent $2^{\alpha-2}$. For example, 5 would be such a number. Indeed,

$$5 = 1 + 4,$$

$$5^2 = 1 + 8 + 16,$$

$$5^4 = 1 + 16 + 32u_2,$$

$$\dotfill$$

$$5^{2^{\alpha-3}} = 1 + 2^{\alpha-2} + 2^\alpha u_{\alpha-3},$$

from which we see that none of the powers 5^1, 5^2, 5^4, ..., $5^{2^{\alpha-3}}$ is congruent to 1 modulo 2^α.

It is not difficult to see that the numbers of the following two rows:

$$5^0,\; 5^1,\; \ldots,\; 5^{2^{\alpha-2}-1},$$

$$-5^0,\; -5^1,\; \ldots,\; -5^{2^{\alpha-2}-1}$$

form a reduced residue system modulo 2^α. Indeed, the number of these numbers is $2 \cdot 2^{\alpha-2} = \varphi(2^\alpha)$; the numbers of each individual row are incongruent among themselves modulo 2^α **(b, §1)**; finally, the numbers of the upper row are incongruent

117

to the numbers of the lower row since the former are congruent
to 1, while the latter are congruent to −1 modulo 4.

e. For convenience in later investigations, we express the
results of b, c, d in more unified form, which is also applicable
in the case $\alpha = 0$.

Let

$$c = 1; \; c_0 = 1, \; \textit{if } \alpha = 0, \; \textit{or } \alpha = 1;$$

$$c = 2; \; c_0 = 2^{\alpha-2}, \; \textit{if } \alpha \geqslant 2$$

*(therefore $cc_0 = \varphi(2^\alpha)$) and let γ and γ_0 run independently
through the least non-negative residues*

$$\gamma = 0, \ldots, c - 1; \; \gamma_0 = 0, \ldots, c_0 - 1$$

*modulo c and c_0. Then $(-1)^\gamma 5^{\gamma_0}$ runs through a reduced
residue system modulo 2^α.*

f. *The congruence*

(1) $$(-1)^\gamma 5^{\gamma_0'} \equiv (-1)^{\gamma'} 5^{\gamma_0'} \pmod{2^\alpha}$$

holds if and only if

$$\gamma \equiv \gamma' \pmod{c}, \; \gamma_0 \equiv \gamma_0' \pmod{c_0}.$$

Indeed, the theorem is evident for $\alpha = 0$. We therefore
assume that $\alpha > 0$. Let the least non-negative residues of
the numbers γ and γ_0 be r and r_0, and of the numbers γ' and
γ_0' be r' and r_0' modulo c and c_0. In view of c, §1 (−1 belongs
to the exponent c, while 5 belongs to the exponent c_0), the
congruence (1) holds if and only if $(1)^r 5^{r_0} \equiv (-1)^{r'} 5^{r_0'} \pmod{2^\alpha}$, i.e. if and only if $r = r'$, $r_0 = r_0'$ (in view of e).

g. If

$$a \equiv (-1)^\gamma 5^{\gamma_0} \pmod{2^\alpha},$$

118

then the system γ, γ_0 is called an *index system of the number a modulo* 2^α.

In view of **e**, every a relatively prime to 2^α (i.e. every odd a) has a unique index system γ', γ_0' in the $cc_0 = \varphi(2^\alpha)$ pairs of values γ, γ_0 considered in **e**.

Knowing a system γ', γ_0', we can also find all index systems of the number a; according to **f**, these will be all pairs γ, γ_0 consisting of the non-negative numbers of the equivalence classes

$$\gamma \equiv \gamma' \pmod{c}, \quad \gamma_0 \equiv \gamma_0' \pmod{c_0}.$$

It follows immediately from the definition we have given of index systems that the numbers with a given index system γ, γ_0 forms an equivalence class of numbers modulo 2^α.

h. *The indices of a product are congruent modulo c and c_0 with the sums of the indices of the factors.*

Indeed, let $\gamma(a)$, $\gamma_0(a)$; $\ldots$; $\gamma(l)$, $\gamma_0(l)$ be index systems for the numbers $a, \ldots, l$. We have

$$a \ldots l \equiv (-1)^{\gamma(a)+\ldots+\gamma(l)} 5^{\gamma_0(a)+\ldots+\gamma_0(l)}$$

Therefore $\gamma(a) + \ldots + \gamma(l)$, $\gamma_0(a) + \ldots + \gamma_0(l)$ are the indices of the product $a \ldots l$.

§7. *Indices for Arbitrary Composite Modulus*

a. Let $m = 2^\alpha p_1^{\alpha_1} p_2^{\alpha_2} \ldots p_k^{\alpha_k}$ be the canonical decomposition of the number m. Moreover let c and c_0 have the values considered in **e**, §6; $c_s = \varphi(p_s^{\alpha_s})$; and let g_s be the smallest primitive root modulo $p_s^{\alpha_s}$.

b. If

(1)
$$a \equiv (-1)^\gamma 5^{\gamma_0} \pmod{2^\alpha},$$

$$a \equiv g_1^{\gamma_1} \pmod{p_1^{\alpha_1}}, \quad \ldots, \quad a \equiv g_k^{\gamma_k} \pmod{p_k^{\alpha_k}},$$

119

then $\gamma, \gamma_0, \gamma_1, \ldots, \gamma_k$ is called an *index system of the number a modulo m*.

It follows from this definition that γ, γ_0 is an index system of the number a modulo 2^α, while $\gamma_1, \ldots, \gamma_k$ are indices of the number a for the moduli $p_1^{\alpha_1}, \ldots, p_k^{\alpha_k}$. Hence (g, §6; c, §4) every a which is relatively prime to m (and hence also relatively prime to all the numbers 2^α, $p_1^{\alpha_1}, \ldots, p_k^{\alpha_k}$), has a unique index system $\gamma', \gamma_0', \gamma_1', \ldots, \gamma_k'$ in the $c c_0 c_1 \ldots c_k = \varphi(m)$ systems which are obtained by letting $\gamma, \gamma_0, \gamma_1, \ldots, \gamma_k$ run independently through the least non-negative residues for the moduli $c, c_0, c_1, \ldots, c_k$, while all the index systems of the number a are just all the systems $\gamma, \gamma_0, \gamma_1, \ldots, \gamma_k$ consisting of the non-negative numbers of the equivalence classes

$$\gamma \equiv \gamma' \pmod{c}, \quad \gamma_0 \equiv \gamma_0' \pmod{c_0},$$

$$\gamma_1 \equiv \gamma_1' \pmod{c_1}, \quad \ldots, \quad \gamma_k \equiv \gamma_k' \pmod{c_k}.$$

The numbers a with a given index system $\gamma, \gamma_0, \gamma_1, \ldots, \gamma_k$ can be found by solving the system (1), and hence they form an equivalence class of numbers modulo m (**b**, §3, ch. **IV**).

c. Since the indices $\gamma, \gamma_0, \gamma_1, \ldots, \gamma_k$ of the number a modulo m are the indices for the respective moduli 2^α, $p_1^{\alpha_1}$, $\ldots, p_k^{\alpha_k}$, we have the theorem:

The indices of a product are congruent modulo $c, c_0, \ldots, c_k$ to the sums of the indices of the factors.

d. Let $r = \varphi(2^\alpha)$ for $\alpha \leqslant 2$ and $r = \dfrac{1}{2}\varphi(2^\alpha)$ for $\alpha > 2$

and let h be the least common multiple of the numbers $r, c_1, \ldots, c_k$. For every a which is relatively prime to m the congruence $a^h \equiv 1$ holds for all the moduli 2^α, $p_1^{\alpha_1}, \ldots, p_k^{\alpha_k}$, which means that this congruence also holds for the modulus m. Hence a cannot be a primitive root modulo m in those cases in which $h < \varphi(m)$. But the latter holds for $\alpha > 2$, for $k > 1$, and for $\alpha = 2$, $k = 1$. Hence for $m > 1$, primitive roots can only exist if $m = 2, 4, p_1^{\alpha_1}, 2p_1^{\alpha_1}$. But the existence of primitive roots in these cases was proven above (§6, §2).

120

Hence

All the cases in which primitive roots modulo m, exceeding 1, exist are just the cases in which

$$m = 2, 4, p^\alpha, 2p^\alpha.$$

Problems for Chapter VI

The letter p always denotes an odd prime, except in problem **11, b** where we also allow the value 2.

1, a. Let a be an integer, $a > 1$. Prove the odd prime divisors of the number $a^p - 1$ divide $a - 1$ or are of the form $2px + 1$.

b. Let a be an integer, $a > 1$. Prove that the odd prime divisors of the number $a^p + 1$ divide $a + 1$ or are of the form $2px + 1$.

c. Prove that there are an infinite number of primes of the form $2px + 1$.

d. Let n be a positive integer. Prove that the prime divisors of the number $2^{2^n} + 1$ are of the form $2^{n+1}x + 1$.

2. Let a be an integer, $a > 1$, and let n be a positive integer. Prove that $\varphi(a^n - 1)$ is a multiple of n.

3, a. Let n be an integer, $n > 1$. Starting from the sequence $1, 2, \ldots, n$ we form, for odd n, the permutations

$$1, 3, 5, \ldots, n - 2, n, n - 1, n - 3, \ldots, 4, 2;$$

$$1, 5, 9, \ldots, 7, 3$$

etc., while for even n we form the permutations

$$1, 3, 5, \ldots, n - 1, n, n - 2, \ldots, 4, 2;$$

$$1, 5, 9, \ldots, 7, 3,$$

etc. Prove that the k-th operation gives the original sequence if and only if $2^k \equiv \pm 1 \pmod{2n - 1}$.

b. Let n be an integer, $n > 1$, and let m be an integer, $m > 1$. We consider the numbers $1, 2, \ldots, n$ in direct order

121

from 1 to n, then in reverse order from n to 2, then in direct order from 1 to n, then in reverse order from n to 2, etc. From this sequence we take the 1-st, $(m + 1)$-st, $(2m + 1)$-st, etc., until we obtain n numbers. We repeat the same operation with this new sequence of n numbers, etc. Prove that the k-th operation gives the original sequence if and only if

$$m^k \equiv \pm 1 \pmod{2n - 1}$$

4. Prove that there exist $\varphi(\delta)$ numbers belonging to the index δ, by considering the congruence $x^\delta \equiv 1 \pmod{p}$ (problem **10, c, ch. IV**) and applying **d, §3, ch. II**.

5, a. Prove that 3 is a primitive root of any prime of the form $2^n + 1$, $n > 1$.

b. Prove that 2 is a primitive root of any prime of the form $2p + 1$ if p is of the form $4n + 1$, while -2 is a primitive root of any prime of the form $2p + 1$ if p is of the form $4n + 3$.

c. Prove that 2 is a primitive root of any prime of the form $4p + 1$.

d. Prove that 3 is a primitive root of any prime of the form

$$2^n p + 1 \text{ for } n > 1 \text{ and } p > \frac{3^{2^{n-1}}}{2^n}.$$

6, a, α) Let n be a positive integer and let $S = 1^n + 2^n + \ldots + (p - 1)^n$. Prove that

$$S \equiv -1 \pmod{p}, \text{ if } n \text{ is a multiple of } p - 1,$$

$$S \equiv 0 \pmod{p}, \text{ otherwise.}$$

β) Using the notation of problem **9, c, ch. V**, prove that

$$S(1) \equiv -\left(\frac{\dfrac{p - 1}{2}}{\dfrac{p - 1}{4}} \right).$$

122

b. Prove Wilson's theorem by applying **b**, §4.

7. Let g and g_1 be primitive roots modulo p, and let $a \operatorname{ind}_g g. \equiv 1 \pmod{p-1}$.

  a. Let $(a, p) = 1$. Prove that

$$\operatorname{ind}_{g_1} a \equiv \alpha \operatorname{ind}_g a \pmod{p-1}.$$

  b. Let n be a divisor of $p - 1$, $1 < n < p - 1$. The numbers relatively prime to p can be divided into n sets by putting those numbers such that $\operatorname{ind} a \equiv s \pmod{n}$ in the s-th set $(s = 0, 1, \ldots, n - 1)$. Prove that that the s-th set for the base g is identical with the s_1-th set for the base g_1, where $s_1 \equiv \alpha s \pmod{n}$.

8. Find the simplest possible method of solving the congruence $x^n \equiv a \pmod{p}$ (convenient for $(n, p - 1)$ not too large) when we know some primitive root g modulo p.

9. Let $m, a, c, c_0, c_1, \ldots, c_k, \gamma, \gamma_0, \gamma_1, \ldots, \gamma_k$ have the values considered in §7. Considering any roots $R, R_0, R_1, \ldots, R_k$ of the equations

$$R^c = 1, \ R_0^{c_0} = 1, \ R_1^{c_1} = 1, \ \ldots, \ R_k^{c_k} = 1,$$

we set

$$\chi(a) = R^\gamma R_0^{\gamma_0} R_1^{\gamma_1} \ldots R_k^{\gamma_k}.$$

If $(a, m) > 1$, then we set $\chi(a) = 0$.

A function defined in this way for all integers a is said to be a *character*. If $R = R_0 = R_1 = \ldots = R_k = 1$, then we say that the character is *principal*; it has the value 1 for $(a, m) = 1$, and the value 0 for $(a, m) > 1$.

  a. Prove that we obtain $\varphi(m)$ different characters in this way (two characters are said to be different if they are not equal for at least one value of a).

  b. Deduce the following properties of characters:

  $\alpha)\ \chi(1) = 1,$

$\beta)$ $\chi(a_1 a_2) = \chi(a_1)\chi(a_2)$,

$\gamma)$ $\chi(a_1) = \chi(a_2)$, if $a_1 \equiv a_2$ (mod m).

c. Prove that

$$\sum_{a=0}^{m-1} \chi(a) = \begin{cases} \varphi(m), \text{ for the principal character,} \\ \\ 0, \text{ for other characters.} \end{cases}$$

d. Prove that, for given a, summing all $\varphi(m)$ characters, we find

$$\sum_{\chi} \chi(a) = \begin{cases} \varphi(m), \text{ if } a \equiv 1 \text{ (mod } m) \\ \\ 0, \quad \text{otherwise.} \end{cases}$$

e. By considering the sum

$$H = \sum_{\chi} \sum_{a} \frac{\chi(a)}{\psi(a)}$$

where a runs through a reduced residue system modulo m, prove that a function $\psi(a)$ defined for all integers a and satisfying the conditions

$$\psi(a) = 0, \text{ if } (a, m) = 1,$$

$$\psi(a) \text{ is not identically equal to } 0,$$

$$\psi(a_1 a_2) = \psi(a_1)\psi(a_2),$$

$$\psi(a_1) = \psi(a_2), \text{ if } a_1 \equiv a_2 \text{ (mod } m),$$

is a character.

f. Prove the following theorems.

$\alpha)$ If $\chi_1(a)$ and $\chi_2(a)$ are characters, then $\chi_1(a)\chi_2(a)$ is also a character.

124

β) If $\chi_1(a)$ is a character and $\chi(a)$ runs through all the characters, then $\chi_1(a)\chi(a)$ also runs through all the characters.

γ) For $(l, m) = 1$, we have

$$\sum_{\chi} \frac{\chi(a)}{\chi(l)} = \begin{cases} \varphi(m), & \text{if } a \equiv l \pmod m \\ \\ 0, & \text{otherwise.} \end{cases}$$

10, a. Let n be a divisor of $p'- 1$, $1 < n \leqslant p - 1$, and let l be an integer which is not divisible by n. The number $R_1 = e^{2\pi i \frac{l}{n}}$ is a root of the equation $R_1^n = 1$, and hence the power $e^{2\pi i \frac{l \text{ ind } x}{n}}$, which is assumed to be equal to 0 for x a multiple of p, is a character modulo p.

α) For $(k, p) = 1$, prove that

$$\sum_{x=1}^{p-1} \exp\left(2\pi i \frac{l \text{ ind } (x + k) - l \text{ ind } x}{n}\right) = -1.$$

β) Let Q be an integer, $1 < Q < p$, and let

$$S = \sum_{x=0}^{p-1} |S_{l,n,x}|^2; \quad S_{l,n,x} = \sum_{z=0}^{Q-1} \exp\left(2\pi i \frac{l \text{ ind } (x + z)}{n}\right)$$

Prove that $S = (p - Q)Q$.

γ) Let M be an integer, $p > 4n^2$, $n > 2$. Prove that the sequence $M, M + 1, \ldots, M + 2[n\sqrt{p}] - 1$ contains a number of the s-th set of problem **7, b.**

b. Let $p > 4\left(\dfrac{p - 1}{\varphi(p\cdot - 1)}\right)^2 2^{2k}$, let k be the number of

different prime divisors of $p - 1$, and let M be an integer.

125

Prove that the sequence $M, M + 1, \ldots, M +$

$+ 2 \left[\dfrac{p - 1}{\varphi(p - 1)} 2^k\sqrt{p} \right] - 1$ contains a primitive root modulo p.

   11, a. Let a be an integer, let n be a divisor of $p - 1$, $1 < n \leqslant p - 1$, and let k be an integer which is not divisible by n,

$$U_{a,p} = \sum_{x=1}^{p-1} \exp\left(2\pi i \frac{k \text{ ind } x}{n}\right) \exp\left(2\pi i \frac{ax}{p}\right)$$

   $\alpha)$ For $(a, p) = 1$, prove that $|U_{a,p}| = \sqrt{p}$.
   $\beta)$ Prove that

$$\exp\left(2\pi i \frac{-k \text{ ind } a}{n}\right) = \frac{U_{a,p}}{U_{1,p}} .$$

   $\gamma)$ Let p be of the form $4m + 1$, and let

$$S = \sum_{x=1}^{p-2} \exp\left(2\pi i \frac{\text{ind } (x^2 + x)}{4}\right).$$

Prove that (cf. problems **9, a** and **9, c, ch. V**) $p = A^2 + B^2$, where A and B are integers defined by the equation $S = A + Bi$.
   b. Let n be an integer, $n > 2$, $m > 1$, $(a, m) = 1$,

$$S_{a,m} = \sum_{x} \exp\left(2\pi i \frac{ax^n}{m}\right) , \quad S'_{a,m} = \sum_{\xi}' \exp\left(2\pi i \frac{a\xi^n}{m}\right) ,$$

where x runs through a complete residue system, while ξ runs through a reduced residue system modulo m (cf. problem **12, d, ch. III** and problem **11, b, ch. V**).
   $\alpha)$ Let $\delta = (n, p - 1)$. Prove that

$$|S_{a,p}| \leqslant (\delta - 1)\sqrt{p} .$$

126

β) Let $(n, p) = 1$ and let s be an integer, $1 < s \leqslant n$. Prove that

$$S_{a, p^s} = p^{s-1}, \quad S'_{a, p^s} = 0.$$

γ) Let s be an integer, $s > n$. Prove that

$$S_{a, p^s} = p^{n-1} S_{a, p^{s-n}}, \quad S'_{a, p^s} = 0.$$

δ) Prove that

$$|S_{a, m}| < Cm^{1-\frac{1}{n}},$$

where C only depends on n.

12. Let M and Q be integers such that $0 < M < M + Q \leqslant p$.

a. Let n be a divisor of $p - 1$, $1 < n < p - 1$, and let k be an integer which is not divisible by n. Prove that

$$\left| \sum_{x=M}^{M+Q-1} \exp\left(2\pi i \frac{k \operatorname{ind} x}{n}\right) \right| < \sqrt{p}\, \ln p.$$

b. Let T be the number of integers of the s-th set of problem **7, b**, contained in the set of numbers M, $M + 1$, $\ldots$, $\ldots$, $M + Q - 1$. Prove that

$$T = \frac{Q}{n} + \theta \sqrt{p}\, \ln p; \quad |\theta| < 1.$$

c. Let k be the number of prime divisors of $p - 1$, and let H be the number of primitive roots modulo p in the set of numbers M, $M + 1$, $\ldots$, $M + Q - 1$. Prove that

$$H = \frac{\varphi(p - 1)}{p - 1} Q + \theta 2^k \sqrt{p}\, \ln p; \quad |\theta| < 1.$$

127

d. Let M_1 and Q_1 be integers, $0 \leqslant M_1 < M_1 + Q_1 \leqslant p - 1$, and let J be the number of integers of the sequence ind M, ind $(M + 1)$, ..., ind $(M + Q - 1)$ in the sequence $M_1, M_1 + 1$, ..., $M_1 + Q - 1$. Prove that

$$J = \frac{QQ_1}{p - 1} + \theta\sqrt{p}\ (\ln p)^2 ; \quad |\theta| < 1.$$

13. Prove that there exists a constant p_0 such that: if $p > p_0$, n is a divisor of $p - 1$, $1 < n < p - 1$, then the smallest of the positive non-residues of degree n modulo p is

$$< h; \quad h = p^{\frac{1}{c}}\ (\ln p)^2 ; \quad c = 2\exp\left(1 - \frac{1}{n}\right).$$

14, a. Let $m > 1$, $(a, m) = 1$,

$$S = \sum_{x=0}^{m-1} \sum_{y=0}^{m-1} \nu(x)\rho(y) \exp\left(2\pi i\,\frac{axy}{m}\right) ;$$

$$\sum_{x=0}^{m-1} |\nu(x)|^2 = X, \quad \sum_{y=0}^{m-1} |\rho(y)|^2 = Y.$$

Prove that $|S| \leqslant \sqrt{XYm}$.

b, α) Let $m > 1$, $(a, m) = 1$, let n be a positive integer, let K be the number of solutions of the congruence $x^n \equiv 1 \pmod{m}$, and let

$$S = \sum_{x=1}^{m-1} \chi(x)e^{2\pi i\,\frac{ax^n}{m}}.$$

Prove that $|S| \leqslant K\sqrt{m}$.

β) Let ϵ be an arbitrary positive constant. For constant n prove that $K = O(m^\epsilon)$ where K is the number considered in problem α).

128

15, a. Let $(a, p) = (b, p) = 1$ and let n be an integer, $|n| = n_1$, $0 < n_1 < p$,

$$S = \sum_{x=1}^{p-1} \exp\left(2\pi i\, \frac{ax^n + bx}{p}\right).$$

Prove that

$$|S| < \frac{3}{2}\, n_1^{\frac{1}{2}}\, p^{\frac{3}{4}}.$$

b. Let $(A, p) = 1$, let n be an integer, $|n| = n_1$, $0 < n_1 < p$, and let M_0 and Q_0 be integers such that $0 < M_0 < M_0 + Q_0 < p$.

α) Let

$$S = \sum_{x=M_0}^{M_0+Q_0-1} \exp\left(2\pi i\, \frac{Ax^n}{p}\right).$$

Prove that $|S| < \dfrac{3}{2}\, n_1^{\frac{1}{2}}\, p^{\frac{3}{4}}\, \ln\, p$.

β) Let M and Q be integers such that $0 < M < M + Q \leqslant p$, and let T be the number of integers of the sequence Ax^n; $x = M_0,\ M_0 + 1,\ \ldots,\ M_0 + Q_0 - 1$, congruent to numbers of the sequence $M,\ M + 1,\ \ldots,\ M + Q_1 - 1$ modulo p.

Prove that

$$T = \frac{Q_1 Q}{p} + \theta\frac{3}{2} n_1^{\frac{1}{2}} p^{\frac{3}{4}}\, (\ln\, p)^2; \quad |\theta| < 1.$$

c. Let b and c be integers, $(a, p) = 1$, $(b^2 - 4ac, p) = 1$.

α) Let γ be an integer,

$$S = \sum_{x=0}^{p-1} \left(\frac{ax^2 + bx + c}{p}\right) \exp\left(2\pi i\, \frac{\gamma x}{p}\right).$$

Prove that $|S| < \dfrac{3}{2} p^{\frac{3}{4}}$.

β) Let M and Q be integers such that $0 < M < M + Q < p$, and let

$$S = \sum_{x=M}^{M+Q-1} \left(\frac{ax^2 + bx + c}{p} \right).$$

Prove that $|S| < \dfrac{3}{2} p^{\frac{3}{4}} \ln p$.

Numerical Exercises for Chapter VI

1, a. Find (in the simplest possible way) the exponent to which 7 belongs modulo 43.

b. Find the exponent to which 5 belongs modulo 108.

2, a. Find the primitive roots modulo 17, 289, 578.

b. Find the primitive roots modulo 41, 1681, 3362.

c. Find the smallest primitive roots modulo:

$$\alpha)\ 1682;\ \beta)\ 3362.$$

3, a. Form the table of indices modulo 17.

b. Form the table of indices modulo 41.

4, a. Find a primitive root modulo 71, using the method of the example of c, §5.

b. Find a primitive root modulo 191.

5, a. Using the table of indices find the number of solutions of the congruences:

$$\alpha)\ x^{60} \equiv 79 \pmod{97};\ \beta)\ x^{55} \equiv 17 \pmod{97};$$

$$\gamma)\ x^{15} \equiv 46 \pmod{97}.$$

b. Find the number of solutions of the congruences:

130

α) $3x^{12} \equiv 31 \pmod{41}$; β) $7x^7 \equiv 11 \pmod{41}$;

γ) $5x^{30} \equiv 37 \pmod{41}$.

6, a. Using the table of indices, solve the congruences:

α) $x^2 \equiv 59 \pmod{67}$; β) $x^{35} \equiv 17 \pmod{67}$;

γ) $x^{30} \equiv 14 \pmod{67}$.

b. Solve the congruences:

α) $23x^5 \equiv 15 \pmod{73}$; β) $37x^6 \equiv 69 \pmod{73}$;

γ) $44x^{21} \equiv 53 \pmod{73}$.

7, a. Using the theorem of **c, §5**, determine the number of solutions of the congruences:

α) $x^3 \equiv 2 \pmod{37}$; β) $x^{16} \equiv 10 \pmod{37}$.

b. Determine the number of solutions of the congruences:

α) $x^5 \equiv 3 \pmod{71}$; β) $x^{21} \equiv 5 \pmod{71}$.

8, a. Applying the methods of problem **8**, solve the congruences (in the solution of the second congruence use the table of primitive roots at the end of the book):

α) $x^7 \equiv 37 \pmod{101}$; β) $x^5 \equiv 44 \pmod{101}$.

b. Solve the congruence

$$x^3 \equiv 23 \pmod{109}.$$

9, a. Using the table of indices, in a reduced residue system modulo 19 find: α) the quadratic residues; β) the cubic residues.

131

b. In a reduced residue system modulo 37, find: α) the residues of degree 15; β) the residues of degree 8.

10, a. In a reduced residue system modulo 43, find: α) the numbers belonging to the exponent 6; β) the primitive roots.

b. In a reduced residue system modulo 61, find: α) the numbers belonging to the exponent 10; β) the primitive roots.

SOLUTIONS OF THE PROBLEMS

Solutions of the Problems for Chapter I.

1. The remainder resulting from the division of $ax + by$ by d, being of the form $ax' + by'$ and less than d, must be equal to zero. Therefore d is a divisor of all numbers of the form $ax + by$, and in particular is a common divisor of the numbers $a \cdot 1 + b \cdot 0 = a$ and $a \cdot 0 + b \cdot 1 = b$. On the other hand, the expression for d shows that every common divisor of the numbers a and b divides d. Therefore $d = (a, b)$, and hence theorem **1, d,** §2 is valid. The theorems of **e,** §2 are deduced as follows: the smallest positive number of the form $amx + bmy$ is $amx_0 + bmy_0$; the smallest positive number of the form $\dfrac{a}{\delta} x + \dfrac{b}{\delta} y$ is

$$\frac{a}{\delta} x_0 + \frac{b}{\delta} y_0 \, .$$

The generalization of these results is trivial.

2. We first note that the difference of two unequal rational fractions $\dfrac{k}{l}$ and $\dfrac{m}{n}$ ($l > 0$, $n > 0$) is numerically $\geqslant \dfrac{1}{ln}$.

We restrict ourselves by the assumption $\delta_s < \delta_{s+1}$. Let $\dfrac{a}{b}$ be an irreducible fraction, which is not equal to δ_s, such that

133

$0 < b \leqslant Q_s$. We cannot have $\delta_s < \dfrac{a}{b} < \delta_{s+1}$; otherwise we would have

$$\frac{a}{b} - \delta_s \geqslant \frac{1}{bQ_s}$$

$$\delta_{s+1} - \frac{a}{b} \geqslant \frac{1}{bQ_{s+1}}$$

$$\delta_{s+1} - \delta_s > \frac{1}{Q_s Q_{s+1}}$$

Therefore $\dfrac{a}{b} < \delta_s$ or $\delta_{s+1} < \dfrac{a}{b}$. In both cases δ_s is closer to α than $\dfrac{a}{b}$.

3. For $n \leqslant 6$ the theorem is evident; we therefore assume $n > 6$. We have

$$\xi = \frac{1 + \sqrt{5}}{2} = 1.618 \ldots; \quad \log_{10}\xi = 0.2 \ldots;$$

$Q_2 > 1 = g_1 = 1$

$Q_3 > Q_2 + 1 > g_2 = 22 > \xi,$

$Q_4 > Q_3 + Q_2 > g_3 = g_2 + g_1 > \xi + 1 = \xi^2,$

$\cdots\cdots\cdots\cdots\cdots\cdots\cdots\cdots\cdots\cdots\cdots\cdots\cdots\cdots\cdots\cdots$

$Q_n > Q_{n-1} + Q_{n-2} > g_{n-1} = g_{n-2} + g_{n-3} > \xi^{n-3} + \xi^{n-4} = \xi^{n-2}.$

Hence

$$N > \xi^{n-2}; \quad n < \frac{\log_{10}N}{\log_{10}\xi} + 2 < 5k + 2; \quad n \leqslant 5k + 1.$$

4, a. For the fractions $\dfrac{0}{1}$ and $\dfrac{1}{1}$ we have $0 \cdot 1 - 1 \cdot 1 =$

$= -1$. Between the fractions $\dfrac{A}{B}$ and $\dfrac{C}{D}$ with $AD - BC = -1$,

we insert the fraction $\dfrac{A + C}{B + D}$, and hence $A(B + D) -$

$- B(A + C) = (A + C)D - (B + D)C = -1$. Therefore the assertion at the end of the problem is true. The existence of

a fraction $\dfrac{k}{l}$ such that $\dfrac{a}{b} < \dfrac{k}{l} < \dfrac{c}{d}$, $l < r$ is impossible.

Otherwise we would have

$$\frac{k}{l} - \frac{a}{b} \geqslant \frac{1}{lb} ; \frac{c}{d} - \frac{k}{l} \geqslant \frac{1}{ld} ; \frac{c}{d} - \frac{a}{b} \geqslant \frac{b + d}{lbd} > \frac{1}{bd}$$

b. It is evident that it is sufficient to consider the case in

which $0 \leqslant \alpha < 1$. Let $\dfrac{a}{b} \leqslant \alpha < \dfrac{c}{d}$, where $\dfrac{a}{b}$ and $\dfrac{c}{d}$ are

neighboring fractions of the Farey series corresponding to r. There are two possible cases:

$$\frac{a}{b} \leqslant \alpha < \frac{a + c}{b + d} ; \frac{a + c}{b + d} \leqslant \alpha < \frac{c}{d}$$

We therefore have one of the two inequalities

$$\left| \alpha - \frac{a}{b} \right| < \frac{1}{b(b + d)} ; \left| \alpha - \frac{c}{d} \right| \leqslant \frac{1}{d(b + d)}$$

from which the required theorem follows because $b + d > r$.

c. For α irrational, the theorem follows from h, §4, if we

take for $\dfrac{P}{Q}$ the convergent $\dfrac{P_{s-1}}{Q_{s-1}}$, where $Q_{s-1} \leqslant r < Q_s$.

135

In the case of rational $\alpha = \dfrac{a}{b}$, the above argument is only valid for $b > r$. But the theorem is true for $b \leqslant r$, since we can then take the fraction $\dfrac{a}{b}$ itself for $\dfrac{P}{Q}$, setting $\theta = 0$.

5, a. The remainder resulting from the division of an odd prime by 4 is either 1 or 3. The product of numbers of the form $4m + 1$ is of the form $4m + 1$. Therefore the number $4p_1 \ldots p_k - 1$, where the $p_1, \ldots, p_k$ are primes of the form $4m + 3$, has a prime divisor q of the form $4m + 3$. Moreover q is different from the primes $p_1, \ldots, p_k$.

b. The primes greater than 3 are of the form $6m + 1$ or $6m + 5$. The number $6p_1 \ldots p_k - 1$, where the $p_1, \ldots, p_k$ are primes of the form $6m + 5$, has a prime divisor q of the form $6m + 5$. Moreover, q is different from the numbers $p_1, \ldots, p_k$.

6. Let $p_1, \ldots, p_k$ be any k primes, and let N be an integer such that $2 < N$, $(3 \ln N)^k < N$. The number of integers a of the sequence $1, 2, \ldots, N$, whose canonical decomposition is of the form $a = p_1^{\alpha_1} \ldots p_k^{\alpha_k}$, is

$$\leqslant \left(\frac{\ln N}{\ln 2} + 1 \right)^k < (3 \ln N)^k < N$$

since $\alpha_s < \dfrac{\ln N}{\ln 2}$. Therefore there are numbers in the sequence $1, 2, \ldots, N$ whose canonical decomposition contains primes different from $p_1, \ldots, p_k$.

7. For example, we obtain such sequences for

$$M = 2 \cdot 3 \cdots (K + 1)t + 2; \quad t = 1, 2, \ldots$$

8. Taking an integer x_0 such that $f(x) > 1$ and $f'(x) > 0$ for $x \geqslant x_0$, we set $f(x_0) = X$. All the numbers $f(x + Xt)$; $t = 1, 2, \ldots$ are composite (multiples of X).

9, a. If (1) holds, one of the numbers x, y, say x, is even; it follows from

$$\left(\frac{x}{2}\right)^2 = \frac{z+y}{2}\,\frac{z-y}{2}$$

where, clearly, $\left(\dfrac{z+y}{2},\ \dfrac{z-y}{2}\right) = 1$, that there exist positive integers u and v such that

$$\frac{x}{2} = uv, \quad \frac{z+y}{2} = u^2, \quad \frac{z-y}{2} = v^2.$$

This implies the necessity of the condition considered in the problem.

The sufficiency of these conditions is evident.

b. In the solution of this problem all letters denote positive integers. Assume the existence of systems x, y, z such that $x^4 + y^4 = z^2$, $x > 0$, $y > 0$, $z > 0$, $(x, y, z) = 1$, and choose the system with smallest z. Assuming x to be even we find $x^2 = 2uv$, $y^2 = u^2 - v^2$, $u > v \geqslant 1$, $(u, v) = 1$, where v is even (for even u we would have $y^2 = 4N + 1$, $u^2 = 4N_1$, $v^2 = 4N_2 + 1$, $4N + 1 = 4N_1 - 4N_2 - 1$, which is impossible). Hence $u = z_1^2$, $v = 2w^2$, $y^2 + 4w^2 = z_1^4$, $2w^2 = 2u_1v_1$, $u_1 = x_1^2$, $v_1 = y_1^2$, $x_1^4 + y_1^4 = z_1^2$, which is impossible since $z_1 < z$.

It follows from the non-solvability of the equation $x^4 + y^4 = z^2$ that the equation $x^4 + y^4 = t^4$ is not solvable in positive integers x, y, t.

10. Setting $x = \dfrac{k}{l}$; $(k, l) = 1$, we find

$$k^n + a_1 k^{n-1} l + \ldots + a_n l^n = 0.$$

Therefore k^n is a multiple of l and hence $l = 1$.

11, a. Let k be the largest integer such that $2^k \leqslant n$ and let P be the product of all the odd numbers which do not exceed n. The number $2^{k-1}PS$ is a sum, all of whose terms, except $2^{k-1}P\dfrac{1}{2^k}$, are integers.

b. Let k be the largest integer such that $3^k \leqslant 2n + 1$ and let P be the product of all the integers relatively prime to 6 which do not exceed $2n + 1$. The number $3^{k-1}PS$ is a sum, all of whose terms, except $3^{k-1}P\dfrac{1}{3^k}$, are integers.

12. For $n \leqslant 8$, the theorem is immediately verifiable. It is therefore sufficient to assume that the theorem is true for the binomials $a + b$, $(a + b)^2$, ..., $(a + b)^{n-1}$ for $n > 8$, and prove that the theorem holds for $(a + b)^n$. But the coefficients of this binomial, except for the extreme ones, which are equal to 1, are just the numbers

$$\frac{n}{1}\, ,\ \frac{n(n - 1)}{1 \cdot 2}\, ,\ \ldots,\ \frac{n(n - 1)\ \ldots\ 2}{1 \cdot 2\ \ldots\ (n - 1)}$$

A necessary and sufficient condition in order that all these numbers be odd is that the extreme numbers, both equal to n, be odd, and the numbers obtained by deleting the odd factors from the numerators and denominators of the remaining numbers be odd.

But, setting $n = 2n_1 + 1$, these numbers can be represented by the terms of the sequence

$$\frac{n_1}{1}\, ,\ \frac{n_1(n_1 - 1)}{1 \cdot 2}\, ,\ \ldots,\ \frac{n_1(n_1 - 1)\ \ldots\ 2}{1 \cdot 2\ \ldots\ (n_1 - 1)}\, .$$

Since $n_1 < n$, the latter are all odd if and only if n_1 is of the form $2^k - 1$, i.e. if and only if n is of the form $2(2^k - 1) + + 1 = 2^{k+1} - 1$.

138

Solutions of the Problems for Chapter II

1, a. On the ordinate of the point of the curve $y = f(x)$ with abscissa x there are $[f(x)]$ lattice points of our region.

b. The required equation follows from $T_1 + T_2 = T$, where T_1, T_2, T denote the number of lattice points of the regions

$$0 < x < \frac{Q}{2}, \ 0 < y < \frac{P}{Q} x,$$

$$0 < y < \frac{P}{2}, \ 0 < x < \frac{Q}{P} y,$$

$$0 < x < \frac{Q}{2}, \ 0 < y < \frac{P}{2}.$$

c. The required equation follows from

$$T = 1 + 4(T_1 + T_2 + T_3 - T_4),$$

where T_1, T_2, T_3, T_4 denote the number of lattice points of the regions

$$x = 0, \ 0 < y \leqslant r;$$

$$0 < x < \frac{r}{\sqrt{2}}, \ 0 < y \leqslant \sqrt{r^2 - x^2} \ ;$$

$$0 < y \leqslant \frac{r}{\sqrt{2}}, \ 0 < x \leqslant \sqrt{r^2 - y^2} \ ;$$

$$0 < x \leqslant \frac{r}{\sqrt{2}}, \ 0 < y \leqslant \frac{r}{\sqrt{2}} \ .$$

d. The required equation follows from $T = T_1 + T_2 - T_3$, where T_1, T_2, T_3 denote the number of lattice points of the regions

$$0 < x \leqslant \sqrt{n} \ , \ 0 < y \leqslant \frac{n}{x} \ ;$$

$$0 < y \leqslant \sqrt{n} \ , \ 0 < x \leqslant \frac{n}{y} \ ;$$

$$0 < x \leqslant \sqrt{n} \ , \ 0 < y \leqslant \sqrt{n} \ .$$

2. The number of positive integers which do not exceed n is equal to $[n]$. Each of them is uniquely representable in the form xk^m, where k is a positive integer; moreover, to a given

x there correspond $\left[\sqrt[m]{\dfrac{n}{x}} \right]$ numbers of this form.

3. We prove the necessity of our conditions. Let N be an integer, $N > 1$. The number of values x such that $[\alpha x] \leqslant N$ can be represented in the form $\dfrac{N}{\alpha} + \lambda$; $0 \leqslant \lambda \leqslant C$, while the number of values y such that $[\beta y] \leqslant N$ can be represented in the form $\dfrac{N}{\beta} + \lambda_1$; $0 \leqslant \lambda_1 \leqslant C_1$, where C and C_1 do not depend on N. Dividing $\dfrac{N}{\alpha} + \lambda + \dfrac{N}{\beta} + \lambda_1 = N$ by N, and letting $N \to \infty$, we find $\dfrac{1}{\alpha} + \dfrac{1}{\beta} = 1$. The latter equation for rational $\alpha = \dfrac{a}{b}$ $(a > b > 0)$ would give $[\alpha b] = [\beta(a - b)]$.

Let our conditions be satisfied. Let c be a positive integer, and let $x_0 = \dfrac{c}{\alpha} + \xi$ and $y_0 = \dfrac{c}{\beta} + \eta$ be the smallest

140

integers such that $x_0 \geqslant \dfrac{c}{\alpha}$, $y_0 \geqslant \dfrac{c}{\beta}$. Evidently, $[\alpha x] \gtrless c$ for $x \gtrless x_0$ and $[\beta y] \gtrless c$ for $y \gtrless y_0$, $0 < \xi < 1$, $0 < \eta < 1$, $\alpha \xi$ and $\beta \eta$ are irrational. Since $x_0 + y_0 = c + \eta + \xi$, we have $\xi + \eta = 1$, $\dfrac{\alpha \xi}{\alpha} + \dfrac{\beta \eta}{\beta} = 1$; therefore one and only one of the numbers $\alpha \xi$ and $\beta \eta$ is less than 1. Therefore, one and only one of the numbers $[\alpha x_0]$ and $[\beta y_0]$ is equal to c.

4, a. Our differences are equal to

$$\{\alpha x_1\},\ \{\alpha(x_2 - x_1)\},\ \ldots,\ \{\alpha(x_t - x_{t-1})\},\ \{-\alpha x_t\},$$

they are non-negative, their sum is equal to 1, there are $t + 1$ of them; therefore at least one of these differences does not exceed $\dfrac{1}{t+1} < \dfrac{1}{\tau}$, and hence there exists a number smaller than $\dfrac{1}{\tau}$ of the form $\{\pm \alpha Q\}$, where $0 < Q \leqslant \tau$. From $\pm \alpha Q = [\pm \alpha Q] + \{\pm \alpha Q\}$, setting $\pm[\pm \alpha Q] = P$, we find that

$$|\alpha Q - P| < \frac{1}{\tau},\ \left| \alpha - \frac{P}{Q} \right| < \frac{1}{Q\tau}.$$

b. Setting $X_0 = [X]$, $Y_0 = [Y]$, $\ldots$, $Z_0 = [Z]$, we consider the sequence formed by the numbers $\{\alpha x + \beta y + \ldots + \gamma z\}$ and the number 1 arranged in non-decreasing order, assuming that $x, y, \ldots, z$ run through the values:

$$x = 0, 1, \ldots, X_0;\ y = 0, 1, \ldots, Y_0;\ z = 0, 1, \ldots, Z_0.$$

We obtain $(X_0 + 1)(Y_0 + 1) \ldots (Z_0 + 1) + 1$ numbers, from which we obtain $(X_0 + 1)(Y_0 + 1) \ldots (Z_0 + 1)$ differences. At least one of these differences does not exceed

$$\frac{1}{(X_0 + 1)(Y_0 + 1) \ldots (Z_0 + 1)} < \frac{1}{XY \ldots Z}.$$

141

It is easy to obtain the required theorem from this.

5. We have $\alpha = cq + r + \{\alpha\}$; $0 \leqslant r < q$,

$$\left[\frac{[\alpha]}{c}\right] = \left[q + \frac{r}{c}\right] = q, \qquad \left[\frac{\alpha}{c}\right] = \left[q + \frac{r + \{\alpha\}}{c}\right] = q.$$

6, a. We have $[\alpha + \beta + \ldots + \lambda] = [\alpha] + [\beta] + \ldots + [\lambda] = [\{\alpha\} + \{\beta\} + \ldots + \{\lambda\}]$.

b. The prime p divides $n!, a!, \ldots, l!$ to the exact powers

$$\left[\frac{n}{p}\right] + \left[\frac{n}{p^2}\right] + \ldots, \qquad \left[\frac{a}{p}\right] + \left[\frac{a}{p^2}\right] + \ldots,$$

$$\ldots, \qquad \left[\frac{l}{p}\right] + \left[\frac{l}{p^2}\right] + \ldots$$

Moreover

$$\left[\frac{n}{p^s}\right] \geqslant \left[\frac{a}{p^s}\right] + \ldots + \left[\frac{l}{p^s}\right].$$

7. Assuming that there exists a number a with the required properties, we represent it in the form

$$a = q_k p^{k+1} + q_{k-1} p^k + \ldots + q_1 p^2 + q_0 p + q';$$

$$0 < q_k < p, \ 0 \leqslant q_{k-1} < p, \ \ldots,$$

$$0 \leqslant q_1 < p, \ 0 \leqslant q_0 < p, \ 0 \leqslant q' < p.$$

By **b, §1**,

$$h = q_k u_k + q_{k-1} u_{k-1} + \ldots + q_1 u_1 + q_0 u_0.$$

142

Moreover, for any $s = 1, 2, \ldots, m$, we have

$$q_{s-1}u_{s-1} + q_{s-2}u_{s-2} + \ldots + q_1 u_1 + q_0 u_0 < u_s.$$

Therefore our expression for h must coincide completely with the one considered in the problem.

8, a. Letting x_1 be an integer, $Q \leqslant \alpha < \beta \leqslant R$, $x_1 < \alpha < \beta < x_1 + 1$, and integrating by parts, we find

$$-\int_\alpha^\beta f(x)dx = \int_\alpha^\beta \rho'(x)f(x)dx = \rho(\beta)f(\beta) - \rho(\alpha)f(\alpha) -$$

$$- \sigma(\beta)f'(\beta) + \sigma(\alpha)f'(\alpha) + \int_\alpha^\beta \sigma(x)f''(x)dx.$$

In particular, for $Q \leqslant x_1$, $x_1 + 1 \leqslant R$, passing to the limit, we have

$$-\int_{x_1}^{x_1+1} f(x)dx = -\frac{1}{2}f(x_1 + 1) - \frac{1}{2}f(x_1) + \int_{x_1}^{x_1+1} \sigma(x)f''(x)dx.$$

We can then obtain the required formula easily.

b. Rewriting the formula of problem **a** in the form

$$\sum_{Q<x\leqslant R} f(x) = \int^R f(x)dx - \int^Q f(x)dx + \rho(R)f(R) - \rho(Q)f(Q) -$$

$$- \sigma(R)f'(R) + \sigma(Q)f'(Q) + \int_Q^\infty \sigma(x)f''(x)dx - \int_R^\infty \sigma(x)f''(x)dx,$$

we obtain the required formula.

143

c. Applying the result of problem **b**, we find

$$\ln 1 + \ln 2 + \ldots + \ln n = C + n \ln n - n +$$

$$+ \frac{1}{2} \ln n + \int_n^\infty \frac{\sigma(x)}{x^2}\, dx = n \ln n - n + O(\ln n).$$

9, a. α) We have (**b**, §1 and problem **5**)

$$(1) \qquad \ln([n]!) = \sum_{p \leqslant n} \left(\left[\frac{n}{p} \right] + \left[\frac{n}{p^2} \right] \right) + \ldots \; \ln p.$$

The right side represents the sum of the values of the function $\ln p$, extended over the lattice points (p, s, u) with prime p of the region $p > 0$, $s > 0$, $0 < u \leqslant \dfrac{n}{p^s}$. The part of this sum corresponding to given s and u is equal to $\Theta \left(\sqrt[s]{\dfrac{n}{u}} \right)$; the part corresponding to given u is equal to $\psi \left(\dfrac{n}{u} \right)$.

β) Applying the result of problem **α)** for $n \geqslant 2$, we have

$$\ln ([n]!) - 2 \ln \left(\left[\frac{n}{2} \right] ! \right) =$$

$$= \psi(n) - \psi \left(\frac{n}{2} \right) + \psi \left(\frac{n}{3} \right) - \psi \left(\frac{n}{4} \right) + \ldots > \psi(n) - \psi \left(\frac{n}{2} \right).$$

Setting $\left[\dfrac{n}{2} \right] = m$, we then find ($[n] = 2m$, $[n] = 2m + 1$)

144

$$\psi(n) - \psi\left(\frac{n}{2}\right) \leqslant \ln \frac{(2m+1)!}{(m!)^2} \leqslant$$

$$\leqslant \ln \left(2^m \frac{3 \cdot 5 \, \ldots \, (2m+1)}{1 \cdot 2 \, \ldots \, m}\right) \leqslant \ln\left(2^m 3^m\right) < n,$$

$$\psi(n) = \psi(n) - \psi\left(\frac{n}{2}\right) + \psi\left(\frac{n}{2}\right) - \psi\left(\frac{n}{4}\right) +$$

$$+ \psi\left(\frac{n}{4}\right) - \psi\left(\frac{n}{8}\right) + \ldots < n + \frac{n}{2} + \frac{n}{4} + \ldots = 2n.$$

γ) We have (by the solution of problem β) and the result of problem **8, c**)

$$\psi(n) - \psi\left(\frac{n}{2}\right) + \psi\left(\frac{n}{3}\right) - \psi\left(\frac{n}{4}\right) + \ldots = \ln \frac{[n]!}{\left(\left[\frac{n}{2}\right]!\right)^2} =$$

$$= [n]\ln[n] - [n] - 2\left[\frac{n}{2}\right]\ln\left[\frac{n}{2}\right] + 2\left[\frac{n}{2}\right] + O(\ln n) =$$

$$= n \ln 2 + O(\ln n).$$

Moreover, for $s \geqslant 2$ we find (problem β))

$$\Theta(\sqrt[s]{n}) - \Theta\left(\sqrt[s]{\frac{n}{2}}\right) +$$

$$+ \Theta\left(\sqrt[s]{\frac{n}{3}}\right) - \ldots \begin{cases} < 2\sqrt{n} \quad \text{always} \\ \\ = 0 \text{ for } s > r; \, r = \left[\dfrac{\ln n}{\ln 2}\right] \end{cases}.$$

145

Therefore

$$0 \leqslant \psi(n) - \psi\left(\frac{n}{2}\right) + \psi\left(\frac{n}{3}\right) - \psi\left(\frac{n}{4}\right) +$$

$$+ \ldots - \left(\Theta(n) - \Theta\left(\frac{n}{2}\right) + \Theta\left(\frac{n}{3}\right) - \Theta\left(\frac{n}{4}\right) + \ldots\right) <$$

$$< 2\sqrt{n} + 2\sqrt[3]{n} + 2\sqrt[4]{n} +$$

$$+ \ldots + 2\sqrt[7]{n} < 2(\sqrt{n} + r\sqrt[3]{n}) = O(\sqrt{n}).$$

b. The result follows from equation (1), the inequality of problem **a**, β) and the equation of problem 8, **c.**

c. The equation of problem **b** for sufficiently large m gives

$$\sum_{m < p \leqslant m^2} \frac{\ln p}{p} = \ln m + O(1) \geqslant \frac{\ln m}{2}, \quad \sum_{m < p \leqslant m^2} \frac{4}{p} > 1.$$

If $p_{n+1} > p_n(1 + \epsilon)$ for all pairs p_n, p_{n+1} such that $m < p_n < p_{n+1} \leqslant m^2$ then we would have

$$\sum_{r=0}^{\infty} \frac{4}{m(1 + \epsilon)^r} > 1$$

which is impossible for sufficiently large m.

d. It is evidently sufficient to consider the case in which n is an integer.

Setting $y(r) = \dfrac{\ln r}{r}$ for r prime and $y(r) = 0$ for $r = 1$, and for r composite, we have (problem **b**)

$$y(1) + y(2) + \ldots + y(r) = \ln r + \alpha(r); \quad |\alpha(r)| < C_1,$$

146

where C_1 is a constant. Hence, for $r > 1$ (we consider $\alpha(1) = 1$)

$$\gamma(r) = \ln r - \ln (r - 1) + \alpha(r) - \alpha(r - 1),$$

$$\sum_{0 < p \leqslant n} \frac{1}{p} = T_1 + T_2; \; T_1 = \sum_{1 < r \leqslant n} \frac{\ln r - \ln (r - 1)}{\ln r},$$

$$T_2 = \sum_{1 < r \leqslant n} \frac{\alpha(r) - \alpha(r - 1)}{\ln n}.$$

We have (8, b)

$$T = \sum_{1 < r \leqslant n} \frac{1}{r \ln r} + \sum_{1 < r \leqslant n} \left(\frac{1}{2r^2 \ln r} + \frac{1}{3r^2 \ln r} + \ldots \right) =$$

$$= C_2 + \ln \ln n + 0 \left(\frac{1}{\ln n} \right),$$

where C_2 is a constant. Moreover we find

$$T = \alpha(2) \left(\frac{1}{\ln 2} - \frac{1}{\ln 3} \right) +$$

$$+ \ldots + \alpha(n - 1) \left(\frac{1}{\ln (n - 1)} - \frac{1}{\ln n} \right) + \frac{\alpha(n)}{\ln n}.$$

But, for an integer $m > 1$, we have

$$C_1 \left(\frac{1}{\ln m} - \frac{1}{\ln (m + 1)} \right) +$$

$$+ C_1 \left(\frac{1}{\ln (m + 1)} - \frac{1}{\ln (m + 2)} \right) + \ldots = \frac{C_1}{\ln m}.$$

147

Therefore the series

$$\alpha(2) \left(\frac{1}{\ln 2} - \frac{1}{\ln 3} \right) + \alpha(3) \left(\frac{1}{\ln 3} - \frac{1}{\ln 4} \right) + \cdots$$

converges; therefore, if C_3 is its sum, then

$$T_2 = C_3 + 0 \left(\frac{1}{\ln n} \right) .$$

e. We have

$$\ln \prod_{p \leqslant n} \left(1 - \frac{1}{p} \right) = - \sum_{p \leqslant n} \frac{1}{p} - \sum_{p \leqslant n} \left(\frac{1}{2p^2} + \frac{1}{3p^3} + \cdots \right) =$$

$$= C' - \ln \ln n + 0 \left(\frac{1}{\ln n} \right)$$

where C' is a constant. Setting $C' = \ln C_0$ in the latter equation, we obtain the required equation.

10, a. This result follows from **c, § 2**.

b. Since $\theta(1) = \psi(1) = 1$, the function $\theta(a)$ satisfies condition **1, a, § 2**. Let $a = a_1 a_z$ be one of the decompositions of a into two relatively prime factors. We have

$$\sum_{d_1 \backslash a_1} \sum_{d_2 \backslash a_2} \theta(d_1 d_2) = \psi(a) = \psi(a_1)\psi(a_z) =$$

(1)

$$= \sum_{d_1 \backslash a_1} \sum_{d_2 \backslash a_2} \theta(d_1)\theta(d_2).$$

If condition **2, a, § 2** is satisfied for all products smaller than a, then, for $d_1 d_2 < a$ we have $\theta(d_1 d_2) = \theta(d_1)\theta(d_2)$, and equation (1) gives $\theta(a_1 a_2) = \theta(a_1)\theta(a_2)$, i.e. condition **2, a, § 2** is also

148

satisfied for all products $a_1 a_2$ equal to a. But condition
2, a, $2 is satisfied for the product $1 \cdot 1$ which is equal to 1.
Therefore, it is satisfied for all products.

11, a. Let $m > 1$; for each given x_m dividing a, the indeterminate equation $x_1 \ldots x_{m-1} x_m = a$ has $r_{m-1}\left(\dfrac{a}{x_m}\right)$ solutions. Therefore

$$r_m(a) = \sum_{x_m \backslash a} r_{m-1}\left(\frac{a}{x_m}\right) ;$$

but when x_m runs through all the divisors of the number a, the numbers $d = \dfrac{a}{x_m}$ run through all these same divisors in reverse order. Therefore

$$r_m(a) = \sum_{d \backslash a} r_{m-1}(d).$$

Hence (problem **10, a**), if the theorem is true for the function $r_{m-1}(a)$, then it is also true for the function $r_m(a)$. But the theorem is true for the function $r_1(a) = 1$, and hence it is always true.

b. If $m > 1$ and the theorem is true for the function $r_{m-1}(a)$, then

$$r_m(a) = r_m(p_1) \ldots r_m(p_k) =$$
$$= (r_{m-1}(1) + r_{m-1}(p_1)) \ldots (r_{m-1}(1) + r_{m-1}(p_k)) =$$
$$= (1 + m - 1)^k = m^k .$$

But the theorem is true for the function $r_1(a)$, and hence it is always true.

c. Let $\epsilon = m\epsilon_2$, $\epsilon_2 = 2\eta$, and let $a = p_1^{\alpha_1} \ldots p_k^{\alpha_k}$ be the canonical decomposition of the number a, where $p_1, \ldots, p_k$

149

are arranged in increasing order. For the function $r_2(a) = r(a)$ we have

$$\frac{r(a)}{a^\eta} \leqslant \frac{\alpha_1 + 1}{2^{a_1 \eta}} \frac{\alpha_2 + 1}{3^{a_2 \eta}} \cdots \frac{\alpha_k + 1}{(k + 1)^{a_k \eta}} \, .$$

Each of the factors of the product on the right is smaller than $\frac{1}{\eta}$; the factors $\frac{\alpha_{r-1} + 1}{r^{a_{r-1} \eta}}$ such that $r > 2^{\frac{1}{\eta}}$ is smaller than

$\frac{\alpha_{r-1} + 1}{2^{a_{r-1}}} \leqslant 1$. Therefore, setting $C = \left(\frac{1}{\eta}\right)^{2^{\frac{1}{\eta}}}$, we find

$$\frac{r(a)}{a^\eta} < C, \quad \lim_{a \to \infty} \frac{r(a)}{a^{\epsilon_2}} \leqslant \lim_{a \to \infty} \frac{C}{a^\eta} = 0.$$

It is evident that $r_m(a) < (r(a))^m$ for $m > 2$. Therefore

$$\lim_{a \to \infty} \frac{r_m(a)}{a^\epsilon} < \lim_{a \to \infty} \left(\frac{r(a)}{a^{\epsilon_2}}\right)^m = 0.$$

    d. We divide the systems of values $x_1, \ldots, x_m$ satisfying our inequality into $[n]$ sets with subscripts $1, 2, \ldots, [n]$. The systems such that $x_1 \ldots x_m = a$ are put in the set with subscript a; the number of these systems is $r_m(a)$.

    12. The series defining $\zeta(s)$ converges absolutely for $R(s) > 1$. Therefore

$$(\zeta(s))^m = \sum_{n_1 = 1}^{\infty} \cdots \sum_{n_m = 1}^{\infty} \frac{1}{(n_1 \ldots n_m)^s}$$

while, for given positive n, the number of systems $n_1, \ldots, n_m$ such that $n_1 \ldots n_m = n$ is equal to $r_m(n)$.

150

13, a. The product $P = \prod_p \dfrac{1}{1 - \dfrac{1}{p^s}}$ converges absolutely

for $R(s) > 1$. Since $\dfrac{1}{1 - \dfrac{1}{p^s}} = \dfrac{1}{p^s} + \dfrac{1}{p^{2s}} + \ldots$ for

$N > 2$, we have

$$\prod_{p \leqslant N} \dfrac{1}{1 - \dfrac{1}{p^s}} = \sum_{0 < n \leqslant N} \dfrac{1}{n^s} + {\sum}' \dfrac{1}{n^s}$$

where the second sum on the right is extended over those numbers n which are not divisible by primes larger than N. As $N \to \infty$, the left side tends to P, the first sum of the right side tends to $\zeta(s)$, while the second sum on the right tends to zero.

b. Let $N > 2$. Assuming that there are no primes other than $p_1, \ldots, p_k$, we find that (cf. the solution of problem **a**)

$$\prod_{j=1}^k \dfrac{1}{1 - \dfrac{1}{p_j}} \geqslant \sum_{0 < n \leqslant N} \dfrac{1}{n} .$$

This inequality is impossible for sufficiently large N because the harmonic series $1 + \dfrac{1}{2} + \dfrac{1}{3} + \ldots$ diverges.

c. Assuming that there are no primes other than $p_1, \ldots, p_k$ we find (problem **a**)

$$\prod_{j=1}^k \dfrac{1}{1 - \dfrac{1}{p_j^2}} = \zeta(2).$$

This equation is impossible in view of the irrationality of

$$\zeta(2) = \frac{\pi^2}{6}.$$

14. The infinite product for $\zeta(s)$ of problem **13, a** converges absolutely for $R(s) > 1$. Therefore

$$\ln \zeta(s) = \sum_p \left(\frac{1}{p^s} + \frac{1}{2p^{2s}} + \frac{1}{3p^{3s}} + \cdots \right)$$

where p runs through all the primes. Differentiating, we find

$$\frac{\zeta'(s)}{\zeta(s)} = \sum_p \left(-\frac{\ln p}{p^s} - \frac{\ln p}{p^{2s}} - \frac{\ln p}{p^{3s}} - \cdots \right) = -\sum_{n=1}^{\infty} \frac{\Lambda(n)}{n^s}.$$

15. Let $N > 2$. Applying theorem **b**, §3, we have

$$\prod_{p \leqslant N} \left(1 - \frac{1}{p^s} \right) = \sum_{0 < n \leqslant N} \frac{\mu(n)}{n^s} + \sum' \frac{\mu(n)}{n^s}$$

where the second sum on the right is extended over those numbers n larger than N which are not divisible by primes exceeding N. Taking the limit as $N \to \infty$, we obtain the required identity.

16, a. We apply **d**, §3 to the case in which

$$\delta = 1, 2, \ldots, [n], \quad f = 1, 1, \ldots, 1.$$

It is then evident that $S' = 1$. Moreover S_d is the number of

values δ which are multiples of d, i.e. $\left[\dfrac{n}{d} \right]$.

b, α) The right side of the equation of problem **a** is the sum of the values of the function $\mu(d)$, extended over the lattice

152

points (d, u) of the region $d > 0$, $0 < u \leqslant \dfrac{n}{d}$. The part of this sum corresponding to a given value of u, is equal to $M\left(\dfrac{n}{u}\right)$.

β) The required equation is obtained by termwise subtraction of the equations

$$M(n) + M\left(\frac{n}{2}\right) + M\left(\frac{n}{3}\right) + M\left(\frac{n}{4}\right) + \ldots = 1,$$

$$2M\left(\frac{n}{2}\right) + \qquad 2M\left(\frac{n}{4}\right) + \ldots = 2.$$

c. Let $n_1 = [n]$; let $\delta_1, \delta_2, \ldots, \delta_n$ be defined by the condition: δ_s is the largest integer whose l-th power divides s, $f_s = 1$. Then $S' = T_{l,n}$, S_d is equal to the number of multiples of d^l not exceeding n, i.e. $S_d = \left[\dfrac{n}{d^l}\right]$. From this we obtain the required expression for $T_{l,n}$.

In particular, since $\pi(2) = \dfrac{\pi^2}{6}$, we have

$$T_{2,n} = \frac{6}{\pi^2} n + O(\sqrt{n}\,)$$

for the number $T_{2,n}$ of integers not exceeding n and not divisible by the square of an integer exceeding 1.

17, a. We obtain the required equation from **d**, §3, if we set

$$\delta_s = (x_s, a), \quad f_s = f(x_s).$$

153

b. We obtain the required equation from **d, §3**, if we set

$$\delta_s = (x_1^{(s)}, \ldots, x_k^{(s)}), \quad f_s = f(x_1^{(s)}, \ldots, x_k^{(s)}).$$

c. Applying **d, §3** to the case

$$\delta = \delta_1, \delta_2, \ldots, \delta_T,$$

$$f = F\left(\frac{a}{\delta_1}\right), \quad F\left(\frac{a}{\delta_2}\right), \quad \ldots, \quad F\left(\frac{a}{\delta_T}\right),$$

where we have written down all the divisors of a in the first row, we have

$$S' = F(a), \quad S_d = \sum_{D \setminus \frac{a}{d}} F\left(\frac{a}{dD}\right) = G\left(\frac{a}{d}\right).$$

d. The required equation follows from

$$P' = f_1^{\sum_{d \setminus \delta_1} \mu(d)} \, f_2^{\sum_{d \setminus \delta_2} \mu(d)} \, \cdots \, f_n^{\sum_{d \setminus \delta_n} \mu(d)}.$$

18, a. We apply the theorem of problem **17, a**, letting x run through the numbers $1, 2, \ldots, a$ and taking $f(x) = x^m$. Then

$$S' = \psi_m(a), \quad S_d = d^m + 2^m d^m + \ldots + \left(\frac{a}{d}\right)^m d^m =$$

$$= d^m \sigma_m \left(\frac{a}{d}\right).$$

b. We have

$$\psi_1(a) = \sum_{d \setminus a} \mu(d) \left(\frac{a^2}{2d} + \frac{a}{2}\right) = \frac{a}{2}\,\varphi(a).$$

154

We can obtain the same result more simply. We first write down the numbers of the sequence $1, \ldots, a$ relatively prime to a in increasing order, and then in decreasing order. The sum of the terms of the two sequences equally distant from the initial terms, is equal to a; the number of terms in each sequence is equal to $\varphi(a)$.

c. We have

$$\psi_2(a) = \sum_{d \backslash a} \mu(d) \left(\frac{a^3}{3\,d} + \frac{a^2}{2} + \frac{a}{6}\,d \right) =$$

$$= \frac{a^2}{3}\,\varphi(a) + \frac{a}{6}\,(1 - p_1) \ldots (1 - p_k).$$

19, a. We apply the theorem of problem **17, a,** letting x run through the numbers $1, 2, \ldots, [z]$ and taking $f(x) = 1$. Then $S' = T_z$, S_d is equal to the number of multiples of d which do not exceed z, i.e. $S_d = \left[\dfrac{z}{d} \right]$.

b. We have

$$T_z = \sum_{d \backslash a} \mu(d) \frac{z}{d} + O(\tau(a)) = \frac{z}{a}\,\varphi(a) + O(a^\epsilon).$$

c. This follows from the equation of problem **19, a.**

20. We apply the theorem of problem **17, a,** letting x run through the numbers $1, 2, \ldots, N$, where $N > a$, and taking $f(x) = \dfrac{1}{x^s}$. We then find

$$\sum_{x \leqslant N} \frac{1}{x^s} = \sum_{d \backslash a} \mu(d) \sum_{0 < x \leqslant \frac{N}{d}} \frac{1}{d^s x^s} = \sum_{d \backslash a} \frac{\mu(d)}{d^s} \sum_{0 < x \leqslant \frac{N}{d}} \frac{1}{x^s}.$$

Taking the limit as $N \to \infty$, we obtain the required identity.

155

21, a. We apply the theorem of problem **17, b**, considering the systems of values $x_1, x_2, \ldots, x_k$ considered in the definition of the probabilities P_N, and taking $f(x_1 x_2, \ldots, x_k) = 1$.

Then $P_N = \dfrac{S'}{N^k}$, $S_d = \left[\dfrac{N}{d}\right]^k$, and we obtain

$$P_N = \frac{\sum\limits_{d=1}^{N} \mu(d) \left[\dfrac{N}{d}\right]^k}{N^k} = \sum_{d=1}^{N} \frac{\mu(d)}{d^k} + O\left(\sum_{d=1}^{N} \frac{1}{Nd^{k-1}}\right).$$

Therefore

$$P_N = (\zeta(k))^{-1} + O(\Delta); \; \Delta = \frac{1}{N} \text{ for } k > 2,$$

$$\Delta = \frac{\ln N}{N} \text{ for } k = 2.$$

b. We have $\zeta(2) = \dfrac{\pi^2}{6}$.

22, a. Elementary arguments show that the number of lattice points (u, v) of the region $u^2 + v^2 \leq \rho^2$; $\rho > 0$, not counting the point $(0, 0)$, is equal to $\pi\rho^2 + O(\rho)$. We apply the theorem of problem **17, b**, considering the coordinates x, y of the lattice points of the region $x^2 + y^2 \leq r^2$, different from $(0, 0)$, and setting $f(x, y) = 1$. Then $T = S' + 1$, S_d is equal to the number of lattice points of the region $u^2 + v^2 \leq \left(\dfrac{r}{d}\right)^2$, not considering the point $(0, 0)$. Therefore

$$S_d = \pi \frac{r^2}{d^2} + O\left(\frac{r}{d}\right),$$

156

$$T = \sum_{d=1}^{[r]} \mu(d)\pi \frac{r^2}{d^2} + O\left(\sum_{d=1}^{[r]} \frac{r}{d}\right) = \frac{6}{\pi}r^2 + O(r \ln r).$$

b. Arguing in analogy to the above, we find

$$T = \sum_{d=1}^{[r]} \mu(d)\frac{4}{3}\pi \frac{r^3}{d^3} + O\left(\sum_{d=1}^{[r]} \frac{r^2}{d^2}\right) = \frac{4\pi r^3}{3\zeta(3)} + O(r^2).$$

23, a. The number of divisors d of the numbers $a = p_1^{a_1} \ldots p_k^{a_k}$ which are not divisible by the square of an integer exceeding 1, and having $\varkappa$ prime divisors, is equal to $\binom{k}{\varkappa}$; moreover $\mu(d) = (-1)^{\varkappa}$. Therefore

$$\sum_{d\backslash a} \mu(d) = \sum_{\varkappa=0}^{k} \binom{k}{\varkappa}(-1)^k = (1 - 1)^k = 0.$$

b. Let a be of the same form as in problem **a.** It is sufficient to consider the case $m < k$. For the sum under consideration we have two expressions

$$\sum \mu(d) = \binom{k}{0} - \binom{k}{1} + \ldots + (-1)^m \binom{k}{m}$$

$$= (-1)^m \left(\binom{k}{m+1} - \binom{k}{m+2} + \ldots\right).$$

If m is even, then for $m \leqslant \dfrac{k}{2}$, the first expression < 0, and for $m > \dfrac{k}{2}$ the second expression is $\geqslant 0$. If m is odd, then

157

for $m \leqslant \dfrac{k}{2}$, the first expression <0, and for $m > \dfrac{k}{2}$ the second expression $\leqslant 0$.

c. The proof is almost the same as in **d**, §3, except that the result of problem **b** must be taken into account.

d. The proof is almost the same as in problems **17, a** and **17, b**.

24. Let d run through the divisors of the number a, let $\Omega(d)$ be the number of prime divisors of the number d, and let $\Omega(a) = s$. Following the process given in the problem, we have

$$(N, q, 1) \leqslant \sum_{\Omega(d) \leqslant m} \mu(d) \left(\frac{N}{q^d} + \theta_d \right) =$$

$$= T + T_0 - T_1; \quad |\theta_d| < 1,$$

$$|T| \leqslant \sum_{\Omega(d) \leqslant m} 1, \quad T_0 = \frac{N}{q} \sum_d \frac{\mu(d)}{d}, \quad |T_1| = \sum_{\Omega(d) > m} \frac{N}{q^d}.$$

Moreover

$$|T| \leqslant \sum_{n=0}^{m} \binom{s}{n} = s^m \leqslant e^{nm} < \exp(5r^{1-\epsilon} \ln r) \frac{qr}{N} \frac{N}{qr} = O(\Delta),$$

$$T_0 = \frac{N}{q} \frac{\displaystyle\prod_{p \leqslant n} \left(1 - \frac{1}{p} \right)}{\displaystyle\prod_{p \backslash q} \left(1 - \frac{1}{p} \right)} = O(\Delta).$$

Finally, letting C_1, C_2, C_3 denote constants, we find

158

$$|T_1| \leqslant \frac{N}{q} \sum_{n=m+1}^{s} \sum_{\Omega(d)=n} \frac{1}{d} \leqslant$$

$$\leqslant \frac{N}{q} \sum_{n=m+1}^{s} \frac{\left(\dfrac{1}{2} + \dfrac{1}{3} + \ldots + \dfrac{1}{p_s}\right)^n}{n!} \leqslant$$

$$\leqslant \frac{N}{q} \sum_{n=m+1}^{s} \left(\dfrac{\dfrac{C_1 + \ln r}{4 \ln r - 1}}{l}\right)^n \leqslant$$

$$\leqslant C_2 \frac{N}{q} \sum_{n=m+1}^{s} \left(\frac{3}{4}\right)^n < C_3 \frac{N}{q} r^{-4\ln \frac{4}{3}} = O(\Delta).$$

25. To every divisor d_1 of the number a such that $d_1 < \sqrt{a}$ there corresponds a divisor d_2 such that $d_2 > \sqrt{a}$, $d_1 d_2 = a$. Here $\mu(d_1) = \mu(d_2)$. Therefore

$$2\sum_{d_1} \mu(d_1) = \sum_{d_1} \mu(d_1) + \sum_{d_2} \mu(d_2) = \sum_{d \backslash a} \mu(d) = 0.$$

26. We consider pairs of numbers d which are not divisible by the square of an integer exceeding 1, and satisfying the condition $\varphi(d) = k$, such that each pair consists of some odd number d_1 and the even number $2d_1$. We have $\mu(d_1) + \mu(2d_1) = 0$.

27. Let $p_1, \ldots, p_k$ are distinct prime numbers. Setting $a = p_1 \ldots p_k$, we have

$$\varphi(a) = (p_1 - 1) \ldots (p_k - 1).$$

If there were no primes other than $p_1, \ldots, p_k$, we would have $\varphi(a) = 1$.

28, a. Our numbers are among the numbers $s\delta$; $s = 1, 2,$
$\ldots, \dfrac{a}{\delta}$. But $(s\delta, a) = \delta$ if and only if $\left(s, \dfrac{a}{\delta}\right) = 1$
(e, §2, ch. I). Therefore the assertion in the problem is true and we have

$$a = \sum_{d \setminus a} \varphi\left(\frac{a}{d}\right) = \sum_{d \setminus a} \varphi(d).$$

b, α) Let $a = p_1^{\alpha_1} \ldots p_n^{\alpha_n}$ be the canonical decomposition of the number a. By **a**, the function $\varphi(a)$ is multiplicative, while

$$p_s^{\alpha_s} = \sum_{d \setminus p^s} \varphi(d), \quad p_s^{\alpha_s - 1} = \sum_{d \setminus p^{s-1}} \varphi(d), \quad p_s^{\alpha_s} - p_s^{\alpha_s - 1} = \varphi(p_s^{\alpha_s}).$$

β) For a positive integer m, we have

$$m = \sum_{d \setminus m} (d).$$

Therefore

$$\varphi(a) = \sum_{d \setminus a} \mu(d) \frac{a}{d}.$$

29. We have (p runs through all the primes)

$$\sum_{n=1}^{\infty} \frac{\varphi(n)}{n^s} = \prod_p \left(1 + \frac{\varphi(p)}{p^s} + \frac{\varphi(p^2)}{p^{2s}} + \ldots\right) =$$

$$= \prod_p \frac{1 - \dfrac{1}{p^s}}{1 - \dfrac{1}{p^{s-1}}} = \frac{\zeta(s-1)}{\zeta(s)}.$$

160

30. We have

$$\varphi(1) + \varphi(2) + \ldots + \varphi(n) =$$

$$= \sum_{d \backslash 1} \frac{\mu(d)}{d} + 2 \sum_{d \backslash 2} \frac{\mu(d)}{d} + \ldots + n \sum_{d \backslash n} \frac{\mu(d)}{d} =$$

$$= \sum_{d=1}^{n} \mu(d) \left(1 + 2 + \ldots + \left[\frac{n}{d} \right] \right) =$$

$$= \sum_{d=1}^{n} \mu(d) \frac{n^2}{2d^2} + O(n \ln n) =$$

$$= \frac{n^2}{2} \sum_{d=1}^{\infty} \frac{\mu(d)}{d^2} + O(n \ln n) = \frac{3}{\pi^2} n^2 + O(n \ln n).$$

Solutions of the Problems for Chapter III

1, a. It follows from

$$P = a_n 10^{n-1} + a_{n-1} 10^{n-2} + \ldots + a_1,$$

that

$$P \equiv a_n + a_{n-1} + \ldots + a_1 \ (\text{mod } 9)$$

since $10 \equiv 1 \ (\text{mod } 9)$. Therefore P is a multiple of 3 if and only if the sum of its digits is a multiple of 3; it is a multiple of 9 if and only if this sum is a multiple of 9.

Noting that $10 \equiv -1 \ (\text{mod } 11)$, we have

$$P \equiv (a_1 + a_3 + \ldots) - (a_2 + a_4 + \ldots) \ (\text{mod } 11).$$

161

Therefore P is a multiple of 3 if and only if the sum of its digits in the odd places minus the sum of its digits in the even places is a multiple of 11.

b. It follows from

$$P = b_n 100^{n-1} + b_{n-1} 100^{n-2} + \ldots + b_1$$

that

$$P \equiv (b_1 + b_3 + \ldots) - (b_2 + b_4 + \ldots) \pmod{101}$$

since $100 \equiv -1 \pmod{101}$. Therefore P is a multiple of 101 if and only if $(b_1 + b_3 + \ldots) - (b_2 + b_4 + \ldots)$ is a multiple of 101.

c. It follows from

$$P = c_n 1000^{n-1} + c_{n-1} 1000^{n-2} + \ldots + c_1$$

that

$$P \equiv c_n + c_{n-1} + \ldots + c_1 \pmod{37}$$

since $1000 \equiv 1 \pmod{37}$. Therefore P is a multiple of 37 if and only if $c_n + c_{n-1} + \ldots + c_1$ is a multiple of 37.

Since $1000 \equiv -1 \pmod{7 \cdot 11 \cdot 13}$, we have

$$P \equiv (c_1 + c_3 + \ldots) - (c_2 + c_4 + \ldots) \pmod{7 \cdot 11 \cdot 13}.$$

Therefore P is a multiple of one of the numbers 7, 11, 13 if and only if $(c_1 + c_3 + \ldots) - (c_2 + c_4 + \ldots)$ is a multiple of that number.

2, a. α) When x runs through a complete system of residues modulo m, then $ax + b$ also runs through a complete residue system; the smallest non-negative residue r of the numbers $ax + b$ also runs through the numbers $0, 1, \ldots, m - 1$.

162

Therefore

$$\sum_x \left\{ \frac{ax + b}{m} \right\} = \sum_{r=0}^{m-1} \frac{r}{m} = \frac{1}{2}(m - 1).$$

β) Applying the result of problem **18, b, ch. II**, we find

$$\sum_\xi \left\{ \frac{a\xi}{m} \right\} = \frac{\psi_1(m)}{m} = \frac{1}{2}\varphi(m).$$

b. For $t = 1$, we have $[f(N + m)] - [f(N)] = a$,

$$\sum \delta = \sum_{x=N+1}^{N+m} [f(x)] - \frac{1}{2}[f(N + m)] + \frac{1}{2}[f(N)] - \frac{1}{2} + \frac{1}{2}m =$$

$$= \sum_{x=N+1}^{N+m} f(x) - \sum_{x=N+1}^{N+m} \{f(x)\} - \frac{1}{2}a + \frac{1}{2}(m - 1) = S;$$

and the case in which $t > 1$ also reduces to this case trivially.

c. Let N, M, P_1, P_2 be integers, $M > 0, P_1 > 0, P_2 > 0$.
The trapezoid with vertices $(N, 0), (N, P_1), (N + M, 0)$,
$(N + M, P_2)$ is a special case of the one considered in problem
b. Therefore equation (1) is also valid for it. Equation (1)
can also be obtained easily for such a trapezoid by considering the rectangle with vertices $(N, 0), (N, P + P), (N + M, 0)$,
$(N + M, P + P)$, which is equal to two such trapezoids. For
this rectangle, the equation

$$\sum{}' \delta = S',$$

analogous to equation (1), is evident. Since $\sum{}' \delta = 2\sum \delta$
this implies $S' = 2S$, so that we obtain equation (1).

163

The analogous formula for the triangle considered in the problem follows trivially from this result. But it is of some interest to consider the following derivation: our triangle can be obtained from a certain parallelogram with integral vertices by dividing it into two equal triangles. Let S be the area of the parallelogram and let $T = \sum \delta$ where the sum is extended over all the lattice points of the parallelogram and δ is defined as in problem b. We will have proven the property of the triangle that interests us, if we prove that $S = T$. We consider a square whose side A increases to infinity. The whole plane can be divided into an infinite number of parallelograms of the above type. Let k be the number of parallelograms completely within the interior of the square, and let R be the number of lattice points in the square. As $A \to \infty$, we find

$$\lim \frac{kS}{A^2} = 1, \ \lim \frac{A^2}{R} = 1, \ \lim \frac{R}{kT} = 1.$$

Multiplying these expressions termwise, we find

$$\lim \frac{S}{T} = 1, \ S = T.$$

3, a. Let r be the smallest positive residue of the number $ax + [c]$ modulo m. We have

$$S = \sum_{r=0}^{m-1} \left\{ \frac{r + \Phi(r)}{m} \right\}$$

where $\epsilon \leqslant \Phi(r) \leqslant \epsilon + h$; $\epsilon = \{c\}$. The theorem is evident for $m \leqslant 2h + 1$. We therefore only consider the case $m > 2h + 1$. Setting

$$\left\{ \frac{r + \Phi(r)}{m} \right\} - \frac{r}{m} = \delta(r),$$

164

we have $-1 + \dfrac{\epsilon}{m} \leqslant \delta(r) < \dfrac{h + \epsilon}{m}$ for $r = m - [h + \epsilon]$,

$\ldots, m - 1$; in the other cases $\dfrac{\epsilon}{m} \leqslant \delta(r) \leqslant \dfrac{h + \epsilon}{m}$.

Therefore

$$-[h + \epsilon] + \epsilon \leqslant S - \frac{m - 1}{2} \leqslant h + \epsilon, \quad \left| S - \frac{1}{2}\, m \right| \leqslant h + \frac{1}{2}.$$

b. We have

$$S = \sum_{z=0}^{m-1} \left\{ \frac{az + \psi(z)}{m} \right\} ; \quad \psi(z) = m(AM + B) + \frac{\lambda}{m}\, z.$$

We apply the theorem of problem **a**, setting $h = |\lambda|$. Then we obtain the required result.

c. We find

$$S = \sum_{z=0}^{m-1} \left\{ f(M) + \frac{az}{m} + \frac{\theta z}{m^2} + \frac{f''(M + z_0)}{2}\, z^2 \right\},$$

$$0 < z_0 < m - 1.$$

We apply the theorem of problem **a**, setting $h = 1 + \dfrac{k}{2}$. We then obtain the required result.

4. We develop A in a continued fraction. Let $Q_n = Q'$ be the largest of the denominators of the convergents which does not exceed m, and note (**h, §4, ch. I**)

$$A = \frac{P'}{Q'} + \frac{\theta'}{Q'm}, \quad (P', Q') = 1, \quad |\theta'| < 1.$$

165

It follows from $m < Q_{n+1} < (q_{n+1} + 1)Q_n \leqslant CQ_n$, where C is a constant which is not larger than all the $q_s + 1$, that, for the largest integer H' such that $H'Q' < m$, we have $H' < C$. Applying the theorem of problem **3, b**, we find

$$\left| \sum_{x=M}^{M+H'Q'-1} \{Ax + B\} - \frac{1}{2}H'Q' \right| \leqslant \frac{3}{2}C.$$

Let $m_1 = m - H'Q'$. If $m_1 > 0$, then, choosing Q'' and H'' depending on m as we chose Q' and H' depending on m, we find

$$\left| \sum_{x=M_1}^{M_1+H''Q''-1} \{Ax + B\} - \frac{1}{2}H''Q'' \right| \leqslant \frac{3}{2}C.$$

Let $m_2 = m_1 - H''Q''$. If $m_2 > 0$, then, as above, we find

$$\left| \sum_{x=M_2}^{M_2+H'''Q'''-1} \{Ax + B\} - \frac{1}{2}H'''Q''' \right| \leqslant \frac{3}{2}C,$$

etc., until we find some $m_k = 0$. We then have $(H'Q' + H''Q'' + \ldots + H^{(k)}Q^{(k)} = m)$

$$\left| \sum_{x=M}^{M+m-1} \{Ax + B\} - \frac{1}{2}m \right| < \frac{3}{2}Ck.$$

The numbers $Q', Q'', \ldots, Q^{(k)}$ satisfy the conditions

$$m \geqslant Q' > m_1 \geqslant Q'' > m_2 \geqslant \ldots > m_{k-1} \geqslant Q^{(k)} \geqslant 1.$$

Therefore (problem **3, ch. I**) $k = O(\ln m)$, and hence the required formula is true.

166

5, a. Let the sum on the left be denoted by S. Let $r = A^{\frac{1}{3}}$. The theorem is evident for $r \leqslant 40$. We therefore assume that $r > 40$. Taking $M_1 = [Q + 1]$, we can find numbers a_1, m_1, θ such that

$$f'(M_1) = \frac{a_1}{m_1} + \frac{\theta_1}{m_1 r}; \quad 0 < m \leqslant r, \quad (a_1, m_1) = 1, \quad |\theta_1| < 1.$$

Taking $M_2 = M_1 + m_1$, we find the numbers a_2, m_2, θ_2 analogously; taking $M_3 = M_2 + m_2$, we find the numbers a_3, m_3, θ_3; etc., until we come to $M_{s+1} = M_s + m_s$ such that $0 \leqslant [R] - M_{s+1} < [r]$. Applying the theorem of problem **3, c,** we find

$$\left| S - \frac{1}{2}(m_1 + m_2 + \ldots + m_s + [R] - M_{s+1}) \right| <$$

$$< s\frac{k + 3}{2} + \frac{1}{2}([R] - M_{s+1}),$$

$$\left| S - \frac{1}{2}(R - Q) \right| < s\frac{k + 3}{2} + \frac{r + 1}{2}.$$

The length of the interval for which $\dfrac{a}{m} - \dfrac{1}{m\,r} \leqslant f'(x) \leqslant$

$\leqslant \dfrac{a}{m} + \dfrac{1}{m\,r}$ does not exceed $\dfrac{2A}{m\,r}$. Therefore there are

$< \dfrac{2A}{m^2 r} + 1$ numbers $m_1, m_2, \ldots, m_s$ associated with the

fraction $\dfrac{a}{m}$. Let a_1 and a_2 be the smallest and largest values

of a associated with a given m.

167

We have

$$\frac{a_2 - a_1}{m} - \frac{2}{m\tau} \leqslant \frac{k(R - Q)}{A} \; ;$$

$$a_2 - a_1 + 1 < \frac{k(R - Q)m}{A} + 1.05.$$

Therefore, there are

$$\left(< \frac{2A}{m^2\tau} + 1 \right) \left(\frac{k(R - Q)m}{A} + 1.05 \right) =$$

$$= \frac{k(R - Q)}{\tau} \left(\frac{2}{m} + \frac{m}{\tau^2} \right) + \left(\frac{2A}{m^2\tau} + 1 \right) 1.05$$

numbers $m_1, m_2, \ldots, m_k$ associated with a given m. Summing the latter expression over all $m = 1, 2, \ldots, [\tau]$, we find

$$S < \frac{k(R - Q)}{\tau} \left(2 \ln \tau + 2 + \frac{\tau^2 + \tau}{2\tau^2} \right) + \frac{10A}{3\tau} 1.05 <$$

$$< \frac{k(R - Q)}{\tau} \ln A + \frac{7}{2} \frac{A}{\tau} \; ,$$

$$\left| S - \frac{1}{2}(R - Q) \right| < 2 \frac{k^2(R - Q)}{\tau} \ln A + 8k \frac{A}{\tau} \; .$$

b. We have

$$\left| \sum_{Q < x \leqslant R} \{ f(x) + 1 - \sigma \} - \frac{1}{2}(R - Q) \right| < \Delta,$$

$$\left| \sum_{Q < x \leqslant R} \{ f(x) \} - \frac{1}{2}(R - Q) \right| < \Delta,$$

168

from which, setting $\delta(x) = \{f(x) + 1 - \sigma\} - \{f(x)\}$, we find

$$\left| \sum_{Q < x \leqslant R} \delta(x) \right| < 2\Delta.$$

But, for $\{f(x)\} < \sigma$ we have $\delta(x) = 1 - \sigma$, while for
$\{f(x)\} \geqslant \sigma$ we have $\delta(x) = -\sigma$, and hence $\left| (1 - \sigma)\psi(\sigma) - \sigma(R - Q - \psi(\sigma)) \right| < 2\Delta$, from which we obtain the required
formula.

6, a. We apply the formula of problem **1, c, ch. II.** Setting
$f(x) = \sqrt{r^2 - x^2}$, we have

$$f'(x) = - \frac{x}{\sqrt{r^2 - x^2}} \, , \quad f''(x) = \frac{-r^2}{(r^2 - x^2)^{3/2}} \, ,$$

$$\frac{1}{r} \leqslant |f''(x)| \leqslant \frac{\sqrt{8}}{r}$$

in the interval $0 \leqslant x \leqslant \dfrac{r}{\sqrt{2}}$. Therefore (problem **8, a, ch. II,**
problem **5, a**)

$$T = 4r + 8 \int_{0}^{\frac{r}{\sqrt{2}}} \sqrt{r^2 - x^2} \, dx + 8\rho \left(\frac{r}{\sqrt{2}} \right) \frac{r}{\sqrt{2}} - 8\rho(0)r -$$

$$- 4\frac{r}{\sqrt{2}} - 4\frac{r^2}{2} + 8\frac{r}{\sqrt{2}} \left\{ \frac{r}{\sqrt{2}} \right\} + O(r^{\frac{2}{3}} \ln r) =$$

$$= \pi r^2 + O(r^{\frac{2}{3}} \ln r).$$

b. We have (problems **11, d** and **1, d, ch. II**)

$$r(1) + r(2) + \ldots + r(n) = 2 \sum_{0 < x \leqslant \sqrt{n}} \left[\frac{n}{x} \right] - [\sqrt{n}\,]^2.$$

169

It is sufficient to consider the case $n > 64$. We divide the interval $X < x \leqslant \sqrt{n}$, where $X = 2n^{\frac{1}{3}}$, into $O(\ln n)$ intervals of the form $M < x \leqslant M'$, where $M' \leqslant 2M$. Setting $f(x) =$

$$= \frac{n}{x}, \text{ we have}$$

$$f'(x) = -\frac{n}{x^2}, \quad f''(x) = \frac{2n}{x^3}; \quad \frac{n}{4M^3} \leqslant f''(x) \leqslant \frac{8n}{4M^3}$$

in the interval $M < x \leqslant M'$. Therefore (problem **5, a**)

$$\sum_{M < x \leqslant M'} \left\{ \frac{n}{x} \right\} = \frac{1}{2}(M' - M) + O(n^{\frac{1}{3}} \ln n),$$

$$\sum_{0 < x \leqslant \sqrt{n}} \left\{ \frac{n}{x} \right\} = \frac{1}{2}\sqrt{n} + O(n^{\frac{1}{3}}(\ln n)^2).$$

Moreover (problem **8, b, ch. II**)

$$\sum_{0 < x \leqslant \sqrt{n}} \frac{n}{x} = En + \frac{1}{2}n \ln n + \rho(\sqrt{n})\sqrt{n} + O(1).$$

Therefore

$$\tau(1) + \tau(2) + \ldots + \tau(n) =$$

$$= 2En + n \ln n + 2\rho(\sqrt{n})\sqrt{n} - \sqrt{n} - n +$$

$$+ 2\sqrt{n}\{\sqrt{n}\} + O(n^{\frac{1}{3}}(\ln n)^2) =$$

$$= n(\ln n + 2E - 1) + O(n^{\frac{1}{3}}(\ln n)^2).$$

7. Let the system be improper and let s be the largest integer such that 2^s enters into an odd number of numbers of the system. We replace one of the latter numbers by the smallest number containing only those powers 2^s which enter into an odd number of integers of the rest of the system.

Let the system be proper. A number smaller than one of the numbers T of this system, differs from T in at least one digit in its representation to the base 2.

8, a. Adding the number $H = 3^n + 3^{n-1} + \ldots + 3 + 1$ to each of the numbers of the system represented in the afore-mentioned manner, we obtain numbers which we can obtain by letting $x_n, x_{n-1}, \ldots, x_1, x_0$ in the same form, run through the values $0, 1, 2$, i.e. we obtain all the values $0, 1, \ldots, 2H$.

b. In this way we obtain $m_1 m_2 \ldots m_k$ numbers which are incongruent to one-another modulo $m_1 m_2 \ldots m_k$, since

$$x_1 + m_1 x_2 + m_1 m_2 x_3 + \ldots + m_1 m_2 \ldots m_{k-1} x_k \equiv$$

$$\equiv x_1' + m_1 x_2' + m_1 m_2 x_3' + \ldots + m_1 m_2 \ldots m_{k-1} x_k'$$

$$(\text{mod } m_1 m_2 \ldots m_k)$$

implies in sequence:

$$x_1 \equiv x_1' \ (\text{mod } m_1), \ x_1 = x_1'; \ m_1 x_2 \equiv m_1 x_2' \ (\text{mod } m_1 m_2), \ x_2 = x_2';$$

$$m_1 m_2 x_3 \equiv m_1 m_2 x_3' \ (\text{mod } m_1 m_2 m_3), \ x_3 = x_3',$$

etc.

9, a. In this way we obtain $m_1 m_2 \ldots m_k$ numbers which are incongruent modulo $m_1 m_2 \ldots m_k$, since

$$M_1 x_1 + M_2 x_2 + \ldots + M_k x_k \equiv M_1 x_1' + M_2 x_2' + \ldots + M_k x_k'$$

$$(\text{mod } m_1 m_2 \ldots m_k)$$

171

would imply (every M_j, different from M_s, is a multiple of m_s)

$$M_s x_s \equiv M_s x_s' \ (\text{mod } m), \quad x_s \equiv x_s' \ (\text{mod } m_s), \quad x_s = x_s'.$$

b. In this way we obtain $\varphi(m_1)\varphi(m_2) \ldots \varphi(m_k) = \varphi(m_1 m_2 \ldots m_k)$ numbers which are incongruent modulo $m_1 m_2 \ldots m_k$ by the theorem of problem **a**, and are relatively prime to $m_1 m_2 \ldots m_k$ since $(M_1 x_1 + M_2 x_2 + \ldots + M_k x_k, m_s) = (M_s x_s, m_s) = 1$.

c. By the theorem of problem **a**, the number $M_1 x_1 + M_2 x_2 + \ldots + M_k x_k$ runs through a complete residue system modulo $m_1 m_2 \ldots m_k$ when $x_1, x_2, \ldots, x_k$ run through complete residue systems modulo $m_1, m_2, \ldots, m_k$. This number is relatively prime to $m_1 m_2 \ldots m_k$ if and only if $(x_1, m_1) = (x_2, m_2) = \ldots = (x_k, m_k) = 1$. Therefore $\varphi(m_1 m_2 \ldots m_k) = \varphi(m_1)\varphi(m_2) \ldots \varphi(m_k)$.

d. To obtain the numbers of the sequence $1, 2, \ldots, p^{\alpha}$ relatively prime to p^{α} we delete the numbers of this sequence which are multiples of p, i.e. the numbers $p, 2p, \ldots, p^{\alpha-1}p$. Therefore $\varphi(p^{\alpha}) = p^{\alpha} - p^{\alpha-1}$. The expression for $\varphi(a)$ follows from the latter and theorem **c**, §4, ch. **II**.

10, a. The first assertion follows from

$$\left\{ \frac{x_1}{m_1} + \ldots + \frac{x_k}{m_k} \right\} = \left\{ \frac{M_1 x_1 + \ldots + M_k x_k}{m} \right\};$$

the second assertion follows from

$$\left\{ \frac{\xi_1}{m_1} + \ldots + \frac{\xi_k}{m_k} \right\} = \left\{ \frac{M_1 \xi_1 + \ldots + M_k \xi_k}{m} \right\}.$$

b. The fractions

$$\left\{ \frac{f_1(x_1, \ldots, w_1)}{m_1} + \ldots + \frac{f_k(x_k, \ldots, w_k)}{m_k} \right\}$$

coincide with the fractions

$$\left\{ \frac{f_1(M_1x_1 + \ldots + M_kx_k, \ldots, M_1w_1 + \ldots + M_kw_k)}{m_1} + \ldots + \right.$$

$$\left. + \frac{f_k(M_1x_1 + \ldots + M_kx_k, \ldots, M_1w_1 + \ldots + M_kw_k)}{m_k} \right\} \,,$$

i.e. with the fractions

$$\left\{ \frac{f_1(x, \ldots, w)}{m_1} + \ldots + \frac{f_k(x, \ldots, w)}{m_k} \right\} \,.$$

The first assertion follows trivially from this. The second assertion is proved analogously.

11, a. If a is a multiple of m, we have

$$\sum_x \exp\left(2\pi i \frac{ax}{m}\right) = \sum_x 1 = m.$$

If a is not a multiple of m, we have

$$\sum_x \exp\left(2\pi i \frac{ax}{m}\right) = \frac{\exp\left(2\pi i \dfrac{am}{m}\right) - 1}{\exp\left(2\pi i \dfrac{a}{m}\right) - 1} = 0.$$

b. For non-integral α, the left side is equal to

$$\left| \frac{\exp(2\pi i\alpha(M + P)) - \exp(2\pi i\alpha M)}{\exp(2\pi i\alpha) - 1} \right| \leqslant \frac{1}{\sin \pi(\alpha)} < \frac{1}{h(\alpha)} \,.$$

173

c. By the theorem of problem **b**, the left side does not exceed T_m, where

$$T_m = \sum \frac{1}{h\left(\dfrac{a}{m}\right)} \, .$$

For odd m,

$$T_m < m \sum_{0 < a < \frac{m}{2}} \ln \frac{2a + 1}{2a - 1} = m \ln m,$$

and for even m,

$$T_m < \frac{m}{2} \sum_{0 < a \leqslant \frac{m}{2}} \ln \frac{2a + 1}{2a - 1} + \frac{m}{2} \sum_{0 < a \leqslant \frac{m}{2}} \ln \frac{2a + 1}{2a - 1} < m \ln m.$$

For $m \geqslant 6$, since $\dfrac{1}{2} - \dfrac{1}{3} = \dfrac{1}{6}$, the bound $m \ln m$ can be decreased to

$$2\frac{m}{6} \sum_{0 < a \leqslant \frac{m}{6}} \ln \frac{2a + 1}{2a - 1} = \frac{m}{3} \ln \left(2 \left[\frac{m}{6} \right] + 1 \right) \, .$$

The latter expression is $> \dfrac{m}{2}$ for $m \geqslant 12$ and $> m$ for $m \geqslant 60$.

12, a. Let $m = p_1^{\alpha_1} \ldots p_k^{\alpha_k}$ be the canonical decomposition of the number m. Setting $p_1^{\alpha_1} = m_1, \ldots, p_k^{\alpha_k} = m_k$, and using the notation of problem **10, a**, we have

$$\sum_{\xi_1} \exp \left(2\pi i \frac{\xi_1}{m_1} \right) \ldots \sum_{\xi_k} \exp \left(2\pi i \frac{\xi_k}{m_k} \right) = \sum_{\xi} \exp \left(2\pi i \frac{\xi}{m} \right) \, .$$

174

For $\alpha_s = 1$, we find

$$\sum_{\xi_s} \exp\left(2\pi i \frac{\xi_s}{m_s}\right) = \sum_{x_s} \exp\left(2\pi i \frac{x_s}{m_s}\right) - 1 = -1.$$

For $\alpha_s > 1$, setting $m_s = p_s m_s'$, we find

$$\sum_{\xi_s} \exp\left(2\pi i \frac{\xi_s}{m_s}\right) =$$

$$= \sum_{x_s} \exp\left(2\pi i \frac{x_s}{m_s}\right) - \sum_{u=0}^{m_s'-1} \exp\left(2\pi i \frac{u}{m_s'}\right) = 0.$$

b. Let m be an integer, $m > 1$. We have $\sum_{x=0}^{m-1} \exp 2\pi i \frac{x}{m} =$
$= 0$. By the theorem of problem **a**, the sum of the terms on the left side of this equation such that $(x, m) = d$, is equal to
$$\mu\left(\frac{m}{d}\right) .$$

c. We find

$$\sum_{\xi} \exp\left(2\pi i \frac{\xi}{m}\right) = \sum_{d \backslash m} \mu(d) S_d ,$$

where, setting $m = m_0 d$, we have

$$S_d = \sum_{u=0}^{m_0-1} \exp\left(2\pi i \frac{u}{m_0}\right) .$$

The latter is equal to 0 for $d < m$ and equal to 1 for $d = m$. From this we obtain the theorem of problem **a**.

175

d. This equation follows from problem **10, b.**

e. We have

$$A(m_1) \ldots A(m_k) = m^{-r} \sum_{a_1} \ldots \sum_{a_k} S_{a_1, m_1} \ldots S_{a_k, m_k},$$

where $a_1, \ldots, a_k$ run through reduced residue systems modulo $m_1, \ldots, m_k$. Hence (problem **d**) the first equation of the problem follows immediately.

We also prove the second result analogously.

13, a. We have

$$\sum_{x=0}^{p-1} \exp\left(2\pi i \frac{nx}{p}\right) = \begin{cases} p, & \text{if } n \text{ is a multiple of } p, \\ 0, & \text{otherwise.} \end{cases}$$

b. Expanding the product corresponding to a given n, we find

$$\sum_{d \backslash a} \frac{\mu(d)}{d} \sum_{x=0}^{d-1} \exp\left(2\pi i \frac{nx}{d}\right).$$

Hence, summing over all the $n = 0, 1, \ldots, a - 1$, we obtain the expression for $\varphi(a)$.

14. The part of the expression on the right corresponding to x dividing a, is equal to

$$\lim_{\epsilon \to 0} 2\epsilon \sum_{k=1}^{\infty} \frac{1}{k^{1+\epsilon}} =$$

$$= \lim_{\epsilon \to 0} \left(2\epsilon\left(\int_1^{\infty} \frac{dx}{x^{1+\epsilon}} + O(1) \right) \right) = 2.$$

Setting $\Phi(K) = \sum_{k=1}^{K} \exp\left(2\pi i \frac{ak}{x}\right)$, the part corresponding to

x, not divisible by a, can be represented in the form

$$\lim_{\epsilon \to 0} 2\epsilon \left(\frac{\Phi(1)}{1} + \frac{\Phi(2) - \Phi(1)}{2^{1+\epsilon}} + \frac{\Phi(3) - \Phi(2)}{3^{1+\epsilon}} + \ldots \right) =$$

$$= \lim_{\epsilon \to 0} 2\epsilon \left(\Phi(1) \left(1 - \frac{1}{2^{1+\epsilon}} \right) + \Phi(2) \left(\frac{1}{2^{1+\epsilon}} - \frac{1}{3^{1+\epsilon}} \right) + \ldots \right).$$

The factor to the right of 2ϵ, is $< x$ in absolute value since $|\Phi(K)| < x$; here $\lim_{\epsilon \to 0} 2\epsilon x = 0$. Therefore the right side of the equation considered in the problem is equal to twice the number of divisors of the number a which are smaller than $\sqrt{a}$, multiplied by δ, i.e. equal to $r(a)$.

15, a. We have

$$(h_1 + h_2)^p =$$

$$= h_1^p + \binom{p}{1} h_1^{p-1} h_2 + \ldots + \binom{p}{p-1} h_1 h_2^{p-1} + h_2^p \equiv$$

$$\equiv h_1^p + h_2^p \pmod{p};$$

$$(h_1 + h_2 + h_3)^p \equiv (h_1 + h_2)^p + h_3^p \equiv h_1^p + h_2^p + h_3^p \pmod{p},$$

etc.

b. Setting $h_1 = h_2 = \ldots = h_a = 1$, the theorem of problem **a** gives Fermat's theorem.

c. Let $(a, p) = 1$. For certain integers $N_1, N_2, \ldots, N_\alpha$ we have

$$a^{p-1} = 1 + N_1 p, \quad a^{p(p-1)} = (1 + N_1 p)^p = 1 + N_2 p^2,$$

$$a^{p^2(p-1)} = 1 + N_3 p^3, \ldots, \quad a^{p^{\alpha-1}(p-1)} = 1 + N_\alpha p^\alpha,$$

$$a^{\varphi(p^\alpha)} \equiv 1 \pmod{p^\alpha}.$$

177

Let $m = p_1^{\alpha_1} \ldots p_k^{\alpha_k}$ be the canonical decomposition of the number m. We have

$$a^{\varphi(p_1^{\alpha_1})} \equiv 1 \pmod{p_1^{\alpha_1}}, \quad \ldots, \quad a^{\varphi(p_k^{\alpha_k})} \equiv 1 \pmod{p_k^{\alpha_k}},$$

$$a^{\varphi(m)} \equiv 1 \pmod{p_1^{\alpha_1}}, \quad \ldots, \quad a^{\varphi(m)} \equiv 1 \pmod{p_k^{\alpha_k}},$$

$$a^{\varphi(m)} \equiv 1 \pmod{m}.$$

Solutions of the Problems for Chapter IV

1, a. The theorem follows immediately from the theorem of problem **11, a, ch. III.**

b. Let d be a divisor of the number m, $m = m_0 d$, and let H_d denote the sum of the terms such that $(a, m) = d$ in the expression for Tm in problem **a.** We find

$$H_d = \sum_{a_0} \sum_{x=0}^{m-1} \ldots \sum_{w=0}^{m-1} \exp\left(2\pi i \frac{a_0 f(x, \ldots, w)}{m_0}\right),$$

where a_0 runs through a reduced residue system modulo m_0. From this we deduce

$$H_d = d^r \sum_{a_0} \sum_{x_0=0}^{m_0-1} \ldots \sum_{w_0=0}^{m_0-1} \exp\left(2\pi i \frac{a_0 f(x_0, \ldots, w_0)}{m_0}\right) = m^r A(m_0).$$

c. Let $m > 0$, $(a, m) = d$, $a = a_0 d$, $m = m_0 d$, and let T be the number of solutions of the congruence $ax \equiv b \pmod{m}$. We have

$$
\begin{aligned}
Tm &= \sum_{a=0}^{m-1} \sum_{x=0}^{m-1} \exp\left(2\pi i \frac{\alpha(ax - b)}{m}\right) \\
&= \sum_{\alpha=0}^{m-1} \sum_{x=0}^{m-1} \exp\left(2\pi i \frac{\alpha a_0}{m_0} x - 2\pi i \frac{b\alpha}{m}\right) \\
&= m \sum_{\alpha_1=0}^{d-1} \exp\left(-2\pi i \frac{b\alpha_1}{d}\right) = \begin{cases} md, & \text{if } b \text{ is a multiple of } d, \\ 0, & \text{otherwise} \end{cases}
\end{aligned}
$$

d. Setting $(a, m) = d_1$, $(b, d_1) = d_2$, ..., $(f, d_{r-1}) = d_r$, $m = d_1 m_1$, $d_1 = d_2 m_2$, ..., $d_{r-1} = d_r m_r$, we find $d = d_r$,

$$Tm = \sum_{a=0}^{m-1} \sum_{x=0}^{m-1} \sum_{y=0}^{m-1} \ldots \sum_{w=0}^{m-1} \exp\left(2\pi i \frac{\alpha(ax + by + \ldots + fw + g)}{m}\right)$$

$$= m \sum_{a_1=0}^{d_1-1} \sum_{y=0}^{m-1} \ldots \sum_{w=0}^{m-1} \exp\left(2\pi i \frac{\alpha_1(by + \ldots + fw + g)}{d_1}\right)$$

....

$$= m^{r-1} \sum_{a_{r-1}=0}^{d_{r-1}-1} \sum_{w=0}^{m-1} \exp\left(2\pi i \frac{\alpha_{r-1}(fw + g)}{d_{r-1}}\right) =$$

$$= m^r \sum_{a_r=0}^{d_r-1} \exp\left(2\pi i \frac{\alpha_r g}{d_r}\right) .$$

e. We apply the method of induction. Using the notation of problem **d**, assume that the theorem is true for r variables. We consider the congruence

(2) $\qquad lv + ax + \ldots + fw + g \equiv 0 \pmod{m}$.

Let $(l, m) = d_0$. Congruence (2) holds if and only if $ax + + \ldots + fw + g \equiv 0 \pmod{d_0}$. The latter congruence holds if and only if g is a multiple of d', where $d' = (a, \ldots, f, d_0) = = (1, a, \ldots, f, m)$, and it has $d_0^{r-1} d'$ solutions. Therefore the congruence (2) holds only if g is a multiple of d', and it then has $d_0^{r-1} d' \left(\dfrac{m}{d_0}\right)^r d_0 = m^r d'$ solutions. Therefore the theorem is also true for $r + 1$ variables. But the theorem is true for one variable, and hence is always true.

2, a. We have $a^{\varphi(m)} \equiv 1 \pmod{m}$, $a \cdot b a^{\varphi(m)-1} \equiv b \pmod{m}$.

179

b. We have

$$1 \cdot 2 \, \ldots \, (a - 1) \, ab \, (-1)^{a-1} \frac{(p - 1) \, \ldots \, (p - a + 1)}{1 \cdot 2 \, \ldots \, (a - 1)} \equiv$$

$$\equiv b \cdot 1 \cdot 2 \, \ldots \, (a - 1) \pmod{p},$$

and dividing by $1 \cdot 2 \, \ldots \, (a - 1)$, we obtain the required theorem.

c. It is evidently sufficient to consider the case $(2, b) = 1$. For an appropriate choice of sign, $b \pm m \equiv 0 \pmod 4$. Let 2^δ be the largest power of 2 dividing $b \pm m$. For $\delta \geqslant k$, we have

$$x \equiv \frac{b \pm m}{2^k} \pmod m.$$

If $\delta < k$, then

$$2^{k - \delta} x \equiv \frac{b \pm m}{2^\delta} \pmod m.$$

We repeat the analogous operation with this congruence, etc.

β) We consider $(3, b) = 1$. For an appropriate choice of sign, we have $b \pm m \equiv 0 \pmod 3$. Let 3^δ be the largest power of 3 dividing $b \pm m$. For $\delta \geqslant k$, we have

$$x \equiv \frac{b \pm m}{2^k} \pmod m.$$

If $\delta < k$, then

$$3^{k - \delta} x \equiv \frac{b \pm m}{3^\delta} \pmod m.$$

We repeat the analogous operation with this congruence, etc.

180

γ) Let p be a prime divisor of the number a. Determine t by the condition $b + mt \equiv 0 \pmod{p}$. Let p^δ be the largest power of p dividing $(a, b + mt)$, and let $a = a_1 p^\delta$. We have

$$a_1 x \equiv \frac{b + mt}{p^\delta} \pmod{m}.$$

If $|a_1| > 1$, then we repeat this operation with the new congruence, etc.

This method is convenient for the case in which a has small prime factors.

3. Setting $t = [r]$, we write the congruences

$$a \cdot 0 \equiv 0 \pmod{m},$$

$$a \cdot 1 \equiv \gamma_1 \pmod{m},$$

$$\dots\dots\dots\dots\dots\dots$$

$$a \cdot t \equiv \gamma_t \pmod{m},$$

$$a \cdot 0 \equiv m \pmod{m}.$$

Arranging these congruences so that their right sides are in order of increase (cf. problem **4, a, ch. II**) and multiplying termwise each congruence (except the last one) by its successor, we obtain $t + 1$ congruences of the form $az \equiv u$ $\pmod{m}$; $0 < |z| \leqslant r$. Here $0 < u < \dfrac{m}{r}$ in at least one congruence. Indeed, u has $t + 1 > r$ values, these values are positive, and their sum is equal to m.

4, a, α) This follows from the definition of symbolic fractions.

β) Here we can set $b_0 = b + mt$, where t is defined by the condition $b + mt \equiv 0 \pmod{a}$; then the congruence $ax \equiv b$ has as solution an integer which represents the ordinary fraction $\dfrac{b_0}{a}$.

γ) We have (b_0 is a multiple of a, d_0 is a multiple of c)

$$\frac{b}{a} + \frac{d}{c} \equiv \frac{b_0}{a} + \frac{d_0}{c} = \frac{b_0 c + a d_0}{ac} \equiv \frac{bc + ad}{ac}.$$

δ) We have

$$\frac{b}{a} \cdot \frac{d}{c} \equiv \frac{b_0}{a} \cdot \frac{d_0}{c} = \frac{b_0 d_0}{ac} \equiv \frac{bd}{ac}.$$

b, α) We have (the congruences are taken modulo p)

$$\binom{p-1}{a} = \frac{(p-1)(p-2) \ldots (p-a)}{1 \cdot 2 \ldots a} \equiv$$

$$\equiv \frac{(-1)^a 1 \cdot 2 \ldots a}{1 \cdot 2 \ldots a} \equiv (-1)^a.$$

Now problem **2, b** is solved more simply as follows:

$$\frac{b}{a} \equiv \frac{b(-1)^{a-1}(p-1) \ldots (p - (a-1))}{1 \cdot 2 \ldots (a-1)^a} \pmod{p}.$$

β) We have

$$\frac{2^p - 2}{p} \equiv 1 + \frac{p-1}{1 \cdot 2} + \frac{(p-1)(p-2)}{1 \cdot 2 \cdot 3} +$$

$$+ \ldots + \frac{(p-1)(p-2) \ldots (p - (p-2))}{1 \cdot 2 \ldots (p-1)} \pmod{p}.$$

5, a. The numbers $s, s + 1, \ldots, s + n - 1$ have no divisors in common with d. The products $s(s+1) \ldots (s + n - 1)$ can be put in $n^\varkappa$ sets in a number of ways equal to the number of ways that d can be decomposed into n relatively prime factors, where order of the factors is taken into account

182

(problem **11, b, ch. II**). Let $d = u_1 u_2 \ldots u_n$ be one of these decompositions. The number of products such that $s \equiv$ $\equiv 0 \pmod{u_1}$, $s + 1 \equiv 0 \pmod{u_2}, \ldots, s + n - 1 \equiv 0 \pmod{u_n}$ is equal to $\dfrac{a}{d}$. Therefore the required number is equal to $n^\varkappa \dfrac{a}{d}$.

b. This number is equal to

$$\sum_{d \setminus a} \mu(d) S_d; \quad S_d = \frac{n^k a}{d},$$

where k is the number of different prime divisors of the number d. But we have

$$\sum_{d \setminus a} \mu(d) \frac{n^k a}{d} = a \left(1 - \frac{n}{p_1}\right) \left(1 - \frac{n}{p_2}\right) \ldots \left(1 - \frac{n}{p_k}\right).$$

6, a. All the values of x satisfying the first congruence are given by the equation $x = b_1 + m_1 t$, where t is an integer. In order to choose from them those values which also satisfy the second congruence, it is only necessary to choose those values of t which satisfy the congruence

$$m_1 t \equiv b_2 - b_1 \pmod{m_2}.$$

But this congruence is solvable if and only if $b_2 - b_1$ is a multiple of d. Moreover, when the congruence is solvable, the set of values of t satisfying it is defined by an equation of the form $t = t_0 + \dfrac{m_2}{d} t'$, where t' is an integer; and hence the set of values of x satisfying the system considered in the problem is defined by the equation

$$x = b_1 + m_1 \left(t_0 + \frac{m_2}{d} t'\right) = x_{1,2} + m_{1,2} t'; \quad x_{1,2} = b_1 + m_1 t_0.$$

183

b. If the system

$$x \equiv b_1 \pmod{m_1}, \quad x \equiv b_2 \pmod{m_2}$$

is solvable, the set of values of x satisfying it is representable in the form $x \equiv x_{1,2} \pmod{m_{1,2}}$. If the system

$$x \equiv x_{1,2} \pmod{m_{1,2}}, \quad x \equiv b_3 \pmod{m_3}$$

is solvable, the set of values of x satisfying it is representable in the form $x \equiv x_{1,2,3} \pmod{m_{1,2,3}}$. If the system

$$x \equiv x_{1,2,3} \pmod{m_{1,2,3}}, \quad x \equiv b_4 \pmod{m_4}$$

is solvable, the set of values of x satisfying it is representable in the form $x \equiv x_{1,2,3,4} \pmod{m_{1,2,3,4}}$, etc.

7, α) If x is replaced by $-x$ (and hence x' is replaced by $-x'$) the sum $\left(\dfrac{a,\, b}{m} \right)$ is not changed.

β) When x runs through a reduced residue system modulo m, x' also runs through a reduced residue system modulo m.

γ) Setting $x \equiv hz \pmod{m}$, we find

$$\left(\frac{a,\, bh}{m} \right) = \sum_z \exp\left(2\pi i \, \frac{ahz + bz'}{m} \right) = \left(\frac{ah,\, b}{m} \right).$$

δ) We have

$$\left(\frac{a_1,\, 1}{m_1} \right) \left(\frac{a_2,\, 1}{m_2} \right) =$$

$$= \sum_x \sum_y \exp\left(2\pi i \, \frac{a_1 m_2 x + a_2 m_1 y + m_2 x' + m_1 y'}{m_1 m_2} \right).$$

184

Setting $m_2 x' + m_1 y' = z'$, we have

$$(a_1 m_2 x + a_2 m_1 y)(m_2 x' + m_1 y') \equiv a_1 m_2^2 + a_2 m_1^2 \pmod{m_1 m_2},$$

$$\left(\frac{a_1,\ 1}{m_1}\right)\left(\frac{a_2,\ 1}{m_2}\right) = \left(\frac{m_2^2 a_1 + m_1^2 a_2,\ 1}{m_1 m_2}\right)$$

which proves our property for the case of two factors. The generalization to the case of more than two factors is trivial.

8. The congruence

$$a_0 x^n + a_1 x^{n-1} + \ldots + a_n - a_0(x - x_1)(x - x_2) \ldots (x - x_n) \equiv$$
$$\equiv 0 \pmod{p}$$

has n solutions. Its degree is less than n. Therefore all its coefficients are multiples of p, and this is also expressed in the congruences considered in the problem.

9, a. Corresponding to x in the sequence $2, 3, \ldots, p - 2$ we find a number x', different from it, in the same sequence such that $xx' \equiv 1 \pmod{p}$; indeed, it would follow from $x = x'$ that $(x - 1)(x + 1) \equiv 0 \pmod{p}$, and hence $x \equiv 1$ or $x \equiv p - 1$. Therefore

$$2 \cdot 3 \ldots (p - 2) \equiv 1 \pmod{p}; \quad 1 \cdot 2 \ldots (p - 1) \equiv -1 \pmod{p}.$$

b. Let $P > 2$. Assuming that P has a divisor u such that $1 < u < P$, we would have $1 \cdot 2 \ldots (P - 1) + 1 \equiv 1 \pmod{u}$.

10, a. We find h such that $a_0 h \equiv 1 \pmod{m}$. The given congruence is equivalent to the following one:

$$x^n + a_1 h x^{n-1} + \ldots + a_n h \equiv 0 \pmod{m}.$$

b. Let $Q(x)$ be the quotient and let $R(x)$ be the remainder resulting from the division of $x^p - x$ by $f(x)$. All the coefficients of $Q(x)$ and $R(x)$ are integers, the degree of $Q(x)$ is $p - n$, the degree of $R(x)$ is less than n,

$$x^p - x = f(x)Q(x) + R(x).$$

185

Let the congruence $f(x) \equiv 0 \pmod p$ have n solutions. These solutions will also be solutions of the congruence $R(x) \equiv$ $\equiv 0 \pmod p$; therefore all the coefficients of $R(x)$ are multiples of p.

Conversely, let all the coefficients of $R(x)$ be multiples of p. Then $f(x)Q(x)$ is a multiple of p for those values of x for which $x^p - x$ is also a multiple of p; therefore the sum of the numbers of solutions of the congruences

$$f(x) \equiv 0 \pmod p, \quad Q(x) \equiv 0 \pmod p$$

is no smaller than p. Let the first have α, and let the second have β solutions. From

$$\alpha \leqslant n, \ \beta \leqslant p - n, \ p - n, \ p \leqslant \alpha + \beta$$

we deduce $\alpha = n$, $\beta = p - n$.

c. Raising the given congruence to the power $\dfrac{p-1}{n}$ termwise, we find that the given condition is necessary. Let this condition be satisfied; it follows from $x^p - x =$

$$= x(x^{p-1} - A^{\frac{p-1}{n}} + A^{\frac{p-1}{n}} - 1)$$ that the remainder resulting

from the division of $x^p - x$ by $x^n - A$ is $(A^{\frac{p-1}{n}} - 1)x$, where

$A^{\frac{p-1}{n}} - 1$ is a multiple of p.

11. It follows from $x_0^n \equiv A \pmod m$, $y^n \equiv 1 \pmod m$ that $(x_0 y)^n \equiv A \pmod m$; here the products $x_0 y$, corresponding to incongruent (modulo m)y, are incongruent. It follows from $x_0^n \equiv A \pmod m$, $x^n \equiv A \pmod m$ that $x^n \equiv x_0^n \pmod m$, while, defining y by the condition $x \equiv y x_0 \pmod m$, we have

$$y^n \equiv 1 \pmod m.$$

186

1. This congruence is equivalent to the following one: $(2ax + b)^2 \equiv b^2 - 4ac \pmod{m}$. Corresponding to each solution $z \equiv z_0 \pmod{m}$ of the congruence $z^2 \equiv b^2 - 4ac \pmod{m}$, from $2ax + b \equiv z_0 \pmod{m}$ we find a solution of the congruence under consideration.

2, a. For $\left(\dfrac{a}{p}\right) = 1$ we have $a^{2m+1} \equiv 1 \pmod{p}$,

$(a^{m+1})^2 \equiv a \pmod{p}$, $x \equiv \pm a^{m+1} \pmod{p}$.

b. For $\left(\dfrac{a}{p}\right) = 1$ we have $a^{4m+2} \equiv 1 \pmod{p}$, $a^{2m+1} \equiv \pm 1$

$\pmod{p}$, $a^{2m+2} \equiv \pm a \pmod{p}$. Since $\left(\dfrac{2}{p}\right) = -1$ we also

have $2^{4m+2} \equiv -1 \pmod{p}$. Therefore, for a certain s, having one of the values $0, 1$, we find

$$a^{2m+2} 2^{(4m+2)} \equiv a \pmod{p}, \quad x \equiv \pm a^{m+2} 2^{(2m+1)s} \pmod{p}.$$

c. Let $p = 2^k h + 1$, where $k \geqslant 3$ and h is odd, $\left(\dfrac{a}{p}\right) = 1$. We have

$$a^{2^{k-1}h} \equiv 1 \pmod{p}, \quad a^{2^{k-2}h} \equiv \pm 1 \pmod{p}, \quad N^{2^{k-1}h} \equiv -1 \pmod{p}.$$

Therefore, for some non-negative integer s_2 we find

$$a^{2^{k-2}h} N^{s_2 2^{k-1}} \equiv 1 \pmod{p}, \quad a^{2^{k-3}h} N^{s_2 2^{k-2}} \equiv \pm 1 \pmod{p};$$

and hence for some non-negative integer s_3 we find

$$a^{2^{k-3}h} N^{s_3 2^{k-2}} \equiv 1 \pmod{p}, \quad a^{2^{k-4}h} N^{s_3 2^{k-3}} \equiv \pm 1 \pmod{p},$$

187

etc.; finally we find

$$a^h N^{2sk} \equiv 1 \ (\text{mod } p), \quad x \equiv \pm a^{\frac{h+1}{2}} N^{sk} \ (\text{mod } p).$$

d. We have

$$1 \cdot 2 \ \ldots \ 2m(p - 2m) \ \ldots \ (p - 2)(p - 1) + 1 \equiv 0 \ (\text{mod } p),$$

$$(1 \cdot 2 \ \ldots \ 2m)^2 + 1 \equiv 0 \ (\text{mod } p).$$

3, a. The conditions for the solvability of congruence (1) and (2) are deduced trivially (**f**, §2 and **k**, §2). The congruence (3) is solvable if and only if $\left(\dfrac{-3}{p} \right) = 1$. But $\left(\dfrac{-3}{p} \right) = \left(\dfrac{p}{3} \right)$, while

$$\left(\frac{p}{3} \right) = \begin{cases} 1, \text{ if } p \text{ is of the form } 6m + 1, \\ -1, \text{ if } p \text{ is of the form } 6m + 5. \end{cases}$$

b. For any distinct primes $p_1, p_2, \ldots, p_k$ of the form $4m + 1$, the smallest prime divisor p of the number $(2p_1 p_2 \ldots p_k)^2 + 1$ is different from $p_1, p_2, \ldots, p_k$, and since $(2p_1 p_2 \ldots p_k)^2 + 1 \equiv 0 \ (\text{mod } p)$, it is of the form $4m + 1$.

c. For any distinct primes $p_1, p_2, \ldots, p_k$ of the form $6m + 1$, the smallest prime divisor p of the number $(2p_1 p_2 \ldots p_k)^2 + 3$ is different from $p_1, p_2, \ldots, p_k$, and since $(2p_1 p_2 \ldots p_k)^2 + 3 \equiv 0 \ (\text{mod } p)$, it is of the form $6m + 1$.

4. There are numbers in the first set which are congruent to $1 \cdot 1, \ 2 \cdot 2, \ \ldots, \ \dfrac{p-1}{2} \cdot \dfrac{p-1}{2}$, i.e. all the quadratic residues of a complete system; a number in the second set is a quadratic non-residue, by definition. But the second set

188

contains with this non-residue, all the products of the non-residue with residues, i.e. it contains all the quadratic non-residues.

5, a. In the number system to the base p, let

$$a = a_{\alpha-1} p^{\alpha-1} + \ldots + a_1 p + a_0$$

and let the required solution (the smallest non-negative residue) be

$$x = x_{\alpha-1} p^{\alpha-1} + \ldots + x_1 p + x_0.$$

We form the table:

$a_{\alpha-1}$	....	a_4	a_3	a_2	a_1	a_0
$2x_0 x_{\alpha-2}$	....	$2x_0 x_4$	$2x_0 x_3$	$2x_0 x_2$	$2x_0 x_1$	x_0^2
$2x_1 x_{\alpha-2}$	....	$2x_1 x_3$	$2x_1 x_2$	x_1^2		
$2x_2 x_{\alpha-3}$	....	x_2^2				
....						

where the column under a_s consists of numbers whose sum is the coefficient of p^s in the decomposition of the square of the right side of (1) in powers of p. We determine x_0 by the condition

$$x_0^2 \equiv a \pmod{p}.$$

Setting $\dfrac{x_0^2 - a_0}{p} = p_1$, we determine x_1 by the condition

$$p_1 + 2x_0 x_1 \equiv a_1 \pmod{p}.$$

Setting $\dfrac{p_1 + 2x_0 x_1 - a_1}{p} = p_2$, we determine x_2 by the

189

condition

$$p_2 + 2x_0x_2 + x_1^2 \equiv a_2 \pmod{p},$$

etc. For given x_0, the numbers $x_1, x_2, \ldots, x_{\alpha-1}$ are uniquely determined since $(x_0, p) = 1$.

b. Here

$$a = a_{\alpha-1} 2^{\alpha-1} + \ldots + a_3 2^3 + a_2 2^2 + a_1 2 + a_0,$$

$$x = x_{\alpha-1} 2^{\alpha-1} + \ldots + x_3 2^3 + x_2 2^2 + x_1 2 + x_0,$$

and we have the following table:

$a_{\alpha-1}$	....	a_4	a_3	a_2	a_1	a_0
$x_0 x_{\alpha-2}$	....	$x_0 x_3$	$x_0 x_2$	$x_0 x_1$		x_0^2
$x_1 x_{\alpha-3}$	....	$x_1 x_2$		x_1^2		
$x_2 x_{\alpha-4}$	....	x_2^2				

We only consider the case $\alpha \geqslant 3$. Since $(a, 2) = 1$, it follows that $a_0 = 1$. Therefore $x_0 = 1$. Moreover $a_1 = 0$, and since $x_0 x_1 + x_1^2 = x_1 + x_1^2 \equiv 0 \pmod{2}$, we must have $a_2 = 0$. For x_1 there are two possible values: 0 and 1. The numbers $x_2, x_3, \ldots, x_{\alpha-2}$ are uniquely determined, while for $x_{\alpha-2}$, there are two possible values: 0 and 1. Therefore, for $\alpha \geqslant 3$, we must have $a \equiv 1 \pmod{8}$, and then the congruence under consideration has 4 solutions.

6. It is evident that P and Q are integers, where Q is congruent modulo p to a number which we obtain by replacing a by z^2, for which it is sufficient to replace $\sqrt{a}$ by z. Therefore $Q \equiv 2^{\alpha-1} z^{\alpha-1} \pmod{p}$; therefore $(Q, p) = 1$ and Q' is determined by the congruence $QQ' \equiv 1 \pmod{p}$. We have

$$P^2 - aQ^2 = (z + \sqrt{a})^\alpha (z - \sqrt{a})^\alpha = (z^2 - a)^\alpha \equiv 0 \pmod{p^\alpha},$$

190

from which it follows that

$$(PQ')^2 \equiv a(QQ')^2 \equiv a \pmod{p^\alpha}.$$

7. Let $m = 2^\alpha p_1^{\alpha_1} \dots p_k^{\alpha_k}$ be the canonical decomposition of the number m. Then m can be represented in the form $m = 2^\alpha ab$, where $(a, b) = 1$, in 2^k ways.

Let $\alpha = 0$. It follows from $(x - 1)(x + 1) \equiv 0 \pmod{m}$, that for certain a and b

$$x \equiv 1 \pmod{a}; \quad x \equiv -1 \pmod{b}.$$

Solving this system, we obtain $x \equiv x_0 \pmod{m}$. Therefore the congruence under consideration has 2^k solutions.

Let $\alpha = 1$. For certain a and b

$$x \equiv 1 \pmod{2a}; \quad x \equiv -1 \pmod{2b}.$$

Solving this system, we obtain $x \equiv x_0 \pmod{m}$. Hence this congruence has 2^k solutions.

Let $\alpha = 2$. For certain a and b

$$x \equiv 1 \pmod{2a}; \quad x \equiv -1 \pmod{2b}.$$

Solving this system, we obtain $x \equiv x_0 \left(\text{mod } \dfrac{m}{2} \right)$. Therefore our congrunece has 2^{k+1} solutions.

Let $\alpha \geqslant 3$. For certain a and b, one of the systems

$$x \equiv 1 \pmod{2a}; \quad x \equiv -1 \pmod{2^{\alpha-1}b}$$

$$x \equiv 1 \pmod{2^{\alpha-1}a}; \quad x \equiv -1 \pmod{2b}$$

is satisfied. Solving one of these systems, we obtain

$x \equiv x_0 \left(\bmod \dfrac{m}{2} \right)$. Therefore our congruence has 2^{k+2}

solutions.

8, a. Defining x' by the congruence $xx' \equiv 1$ (mod p), we have

$$\sum_{x=1}^{p-1} \left(\frac{x(x+k)}{p} \right) = \sum_{x=1}^{p-1} \left(\frac{xx'(xx' + kx')}{p} \right) =$$

$$= \sum_{x=1}^{p-1} \left(\frac{1 + kx'}{p} \right).$$

It is evident that $1 + kx'$ runs through all the residues of a complete system, except 1. The required theorem follows from this.

b. The required equation follows from

$$T = \frac{1}{4} \sum_{x=1}^{p-2} \left(1 + \epsilon \left(\frac{x}{p} \right) \right); \ \left(1 + \eta \left(\frac{x+1}{p} \right) \right) =$$

$$= \frac{1}{4} \sum_{x=1}^{p-2} \left(1 + \epsilon \left(\frac{x}{p} \right) + \eta \left(\frac{x+1}{p} \right) + \epsilon\eta \left(\frac{x(x+1)}{p} \right) \right).$$

c. We have

$$S \leqslant X \sum_{x=0}^{p-1} \sum_{y_1} \sum_{y} \left(\frac{(xy_1 + 5)(xy + 5)}{p} \right).$$

The part of the expression of the right corresponding to the case $y_1 = y$, does not exceed XpY. We consider the part corresponding to a pair of unequal values y_1 and y, where we assume that $y > 0$ for the sake of definiteness. Setting

192

$xy + k \equiv z \pmod{p}$, we reduce this part to the form

$$X \sum_{z=0}^{p-1} \left(\cfrac{z \dfrac{y_1}{y} z + k \left(1 - \dfrac{y_1}{y}\right)}{p} \right)$$

from which (problem **a**) we find that it is $< X$ in absolute value. Therefore $S^2 < XpY + XY^2 \leqslant 2pXY$.

d, α) We have

$$S = \sum_{x=0}^{p-1} \sum_{z_1=0}^{Q-1} \sum_{z=0}^{Q-1} \left(\frac{(x + z_1)(x + z)}{p} \right).$$

For $z_1 = z$, summation with respect to x gives $p - 1$. For $z_1 \neq z$, summation with respect to x (problem **a**) gives -1. Therefore

$$S = (p - 1)Q - Q(Q - 1) = (p - Q)Q.$$

β) By the theorem of problem α) we have

$$T(Q^{0.5+0.5\lambda})^2 < pQ; \quad T < pQ^{-\lambda}.$$

γ) Setting $[\sqrt{p}\,] = Q$, we apply the theorem of problem α). Assuming there are no quadratic non-residues in the sequence under consideration, we find that $|S_x| \geqslant Q - 1$ for $x = M, M + 1, \ldots, M + 2Q - 1$ and hence

$$2Q(Q - 1)^2 \leqslant (p - Q)Q, \quad 2(Q - 1)^2 < (Q + 1)^2 - Q,$$

$$Q^2 - 5Q < 0,$$

which is impossible for $Q \geqslant 5$.

9, a. If m is representable in the form (1), then the solution

(5)
$$z \equiv z_0 \pmod{m}$$

of the congruence $x \equiv zy \pmod{m}$ is also a solution of the congruence (2). We will say that our representation is associated with the solution (5) of the congruence (2).

With each solution (5) of the congruence (2) is associated not less than one representation (1). Indeed, taking $r = \sqrt{m}$, we have

$$\frac{z_0}{m} = \frac{P}{Q} + \frac{\theta}{Q\sqrt{m}}; \quad (P, Q) = 1, \ 0 < Q \leqslant \sqrt{m}, \ |\theta| < 1.$$

Therefore $z_0 Q = mP + r$, where $|r| < \sqrt{m}$. Moreover, it follows from (2) that $|r|^2 + Q^2 \equiv 0 \pmod{m}$. From this and from $0 < |r|^2 + Q^2 < 2m$, we find

(6)
$$m = |r|^2 + Q^2.$$

Here $(|r|, Q) = 1$, since

$$1 = \frac{r^2 + Q^2}{m} = \frac{(z_0 Q - mP)z_0 Q - rmP + Q^2}{m} \equiv rP \pmod{Q}.$$

If $|r| = r$, then the representation (6) is associated with the solution (5) because $r \equiv z_0 Q \pmod{m}$. If $|r| = -r$, then the representation $m = Q^2 + |r|^2$ is associated with the solution (5) because $z_0^2 Q \equiv z_0 r \pmod{m}$, $Q \equiv z_0 |r| \pmod{m}$.

No more than one representation (1) is associated with each solution (5). Indeed, if there were two representations $m = x^2 + y^2$ and $m = x_0^2 + y_0^2$ of the number m in the form (1) associated with a single solution (5), then $x \equiv z_0 y \pmod{m}$, $x_1 \equiv z_0 y_1 \pmod{m}$ would imply that $xy_1 \equiv x_1 y \pmod{m}$. There-

194

fore $xy_1 = x_1y$, from which it follows that $x = x_1$, $y - y_1$ because $(x, y) = (x_1, y_1) = 1$.

b. If m is representable in the form (3), then the solution

$$(7) \qquad\qquad z \equiv z_0 \ (\mathrm{mod}\ p)$$

of the congruence $x \equiv zy$ (mod p) is also a solution of the congruence (4). We will say that this representation is associated with the solution (7) of the congruence (4).

Knowing a solution (7) of the congruence (4), there is no more than one representation (3). Indeed, taking $r = \sqrt{p}$, we have

$$\frac{z_0}{p} = \frac{P}{Q} + \frac{\theta}{Q\sqrt{p}}; \quad (P,\ Q) = 1,\ 0 < Q \leqslant \sqrt{p},\ |\theta| < 1.$$

Therefore $z_0 Q \equiv r$ (mod p), where $|r| < p$. Moreover, it follows from (4) that $|r|^2 + aQ^2 \equiv 0$ (mod p). From this and from $0 < |r|^2 + aQ^2 < (1 + a)p$ it follows that we must have $|r|^2 + 2Q^2 = p$ or $|r|^2 + 2Q^2 = 2p$ for $a = 2$. In the latter case, $|r|$ is even, $|r| = 2r_1$, $p = Q^2 + 2r_1^2$. For $a = 3$ we must have $|r|^2 + 3Q^2 = p$, or $|r|^2 + 3Q^2 = 2p$, or $|r|^2 + 3Q^2 = 3p$. The second case is impossible: modulo 4 the left side is congruent to 0 while the right side is congruent to 2. In the third case, $|r|$ is a multiple of 3, $|r| = 3r_1$, $p = Q^2 + 3r_1^2$.

Assuming that two representations $p = x^2 + ay^2$ and $p = x_1^2 + ay_1^2$ of the number p in the form (3) are associated with a single solution of the congruence (4), we find $x = x_1$, $y = y_1$. Assuming that these representations are associated with different solutions of the congruence (4), we find $x \equiv zy$ (mod p), $x_1 \equiv -zy$ (mod p) and hence $xy_1 + x_1y \equiv 0$ (mod p). which is impossible because

$$0 < (xy_1 + x_1y)^2 < (x^2 + y^2)(x_1^2 + y_1^2) < p.$$

195

c, α) The terms of the sum $S(k)$ with $x = x_1$ and $x = -x_1$ are equal.

β) We have

$$S(kt^2) = \sum_{x=0}^{p-1} \left(\frac{xt(x^2t^2 + kt^2)}{p} \right) = \left(\frac{t}{p} \right) S(k).$$

γ) Setting $p - 1 = 2p_1$, we have

$$p_1(S(r))^2 + p_1(S(n))^2 = \sum_{t=1}^{p_1} (S(rt^2))^2 + \sum_{t=1}^{p_1} (S(nt^2))^2 =$$

$$= \sum_{k=1}^{p-1} (S(k))^2 = \sum_{x=1}^{p-1} \sum_{y=1}^{p-1} \sum_{k=1}^{p-1} \left(\frac{xy(x^2 + k)(y^2 + k)}{p} \right).$$

For y different from x and $p - x$, the result of summation with respect to k is $-2 \left(\dfrac{xy}{p} \right)$; for $y = x$ and $y = p - x$ it is $(p - 2) \left(\dfrac{xy}{p} \right)$. Therefore

$$p_1(S(r))^2 + p_1(S(n))^2 = 4pp_1, \ p = \left(\frac{1}{2}S(r) \right)^2 + \left(\frac{1}{2}S(n) \right)^2 .$$

10, a. We have

$$X^2 - DY^2 =$$

$$= (x_1 + y_1\sqrt{D})(x_2 \pm y_2\sqrt{D})(x_1 - y_1\sqrt{D})(x_2 \mp y_2\sqrt{D}) = k^2.$$

b. Taking any r_1 such that $r_1 > 1$, we find integers x_1, y_1 such that $\left| y_1\sqrt{D} - x_1 \right| < \dfrac{1}{r_1}$, $0 < y \leqslant r_1$, and multiplying this termwise by $y_1\sqrt{D} + x_1 < 2y_1\sqrt{D} + 1$, we find $\left| x_1^2 - Dy_1^2 \right| < 2\sqrt{D} + 1$. Taking $r_2 > r_1$ so that

196

$\left| y_1 \sqrt{D} - x_1 \right| > \dfrac{1}{r_2}$, we find new integers x_2, y_2 such that $\left| x_2^2 - D y_2^2 \right| < 2\sqrt{D} + 1$, etc.

It is evident that there exists an integer k, not equal to zero, in the interval $-2\sqrt{D} - 1 < k < 2\sqrt{D} + 1$ such that there is an infinite set of pairs x, y with $x^2 - D y^2 = k$ among the pairs $x_1, y_1; x_2, y_2; \ldots;$ among these pairs there are two pairs ξ_1, η_1 and ξ_2, η_2 such that $\xi_1 \equiv \xi_2 \pmod{|k|}$, $\eta_1 \equiv \eta_2 \pmod{|k|}$. Defining the integers ξ_0, η_0 by means of the equation $\xi_0 + \eta_0 \sqrt{D} = (\xi_1 + \eta_1 \sqrt{D})(\xi_2 + \eta_2 \sqrt{D})$, we have (problem **a**)

$$\xi_0^2 - D \eta_0^2 = |k|^2; \quad \xi_0 \equiv \xi_1^2 - D\eta_1^2 \equiv 0 \pmod{k};$$

$$\eta_0 \equiv -\xi_1 \eta_1 + \xi_1 \eta_1 \equiv 0 \pmod{|k|}.$$

Therefore $\xi_0 = \xi |k|$, $\eta_0 = \eta |k|$, where ξ and η are integers and $\xi^2 - D\eta^2 = 1$.

c. The numbers x, y defined by the equation (2) satisfy (problem **a**) the equation (1).

Assuming that there exist pairs of integers x, y satisfying equation (1), but different from the pairs determined by the equation (2), we have

$$(x_0 + y_0 \sqrt{D})^r < x + y\sqrt{D} < (x_0 + y_0 \sqrt{D})^{r+1}$$

for certain $r = 1, 2, \ldots$. Dividing this termwise by $(x_0 + y_0 \sqrt{D})^r$, we find

$$(3) \qquad 1 < X + Y\sqrt{D} < x_0 + y_0 \sqrt{D},$$

where (problem **a**) X and Y are integers determined by the equation

$$X + Y\sqrt{D} = \dfrac{x + y\sqrt{D}}{(x_0 + y_0 \sqrt{D})^r} = (x + y\sqrt{D})(x - y\sqrt{D})^r$$

197

and satisfying the equation

(4) $$X^2 - DY^2 = 1.$$

But from (4) we obtain the inequality $0 < |X| - |Y\sqrt{D}| < 1$, which along with the first inequality of (3) shows that X and Y are positive. Therefore the second inequality of (3) contradicts the definition of x_0 and y_0.

11, a, α) We have

$$|U_{a,p}|^2 = U_{a,p}\overline{U}_{a,p} = \sum_{t=1}^{p-1} \sum_{x=1}^{p-1} \left(\frac{t}{p}\right) \exp\left(2\pi i \frac{ax(t-1)}{p}\right) .$$

For $t = 1$, summation with respect to x gives $p - 1$; for

$t > 1$ it gives $-\left(\dfrac{t}{p}\right)$. Therefore

$$|U_{a,p}|^2 = p - 1 - \sum_{t=2}^{p-1} \left(\frac{t}{p}\right) = p, \quad |U_{a,p}| = \sqrt{p} ,$$

or

$$|U_{a,p}|^2 = U_{a,p}\overline{U}_{a,p} = \sum_{t=0}^{p-1} \sum_{x=0}^{p-1} \left(\frac{x+t}{p}\right) \left(\frac{x}{p}\right) \exp\left(2\pi i \frac{at}{p}\right) .$$

For $t = 0$ summation with respect to x gives $p - 1$; for

$t > 0$ it gives $-\exp\left(2\pi i \dfrac{at}{p}\right)$. Therefore

$$|U_{a,p}|^2 = p - 1 - \sum_{t=1}^{p-1} \exp\left(2\pi i \frac{at}{p}\right) = p, \quad |U_{a,p}| = \sqrt{p} .$$

198

β) The theorem is evident for $(a, p) = p$. For $(a, p) = 1$ it follows from

$$U_{a,p} = \frac{a}{p} \sum_{x=1}^{p-1} \left(\frac{ax}{p}\right) \exp\left(2\pi i \frac{ax}{p}\right) = \left(\frac{a}{p}\right) U_{1,p}.$$

b, α) Let r run through the quadratic residues and let n run through the quadratic non-residues, in a complete system of residues. We have

$$S_{a,p} = 1 + 2 \sum \exp\left(2\pi i \frac{ar}{p}\right).$$

Subtracting

$$0 = 1 + \sum_r \exp\left(2\pi i \frac{ar}{p}\right) + \sum_n \exp\left(2\pi i \frac{an}{p}\right)$$

from the latter termwise, we obtain the required equation.

β) We have

$$|S_{a,m}|^2 = \sum_{t=0}^{m-1} \sum_{x=0}^{m-1} \exp\left(2\pi i \frac{a(t^2 + 2tx)}{m}\right).$$

For given t, summation with respect to x gives $m \exp\left(2\pi i \frac{at^2}{m}\right)$ or 0 according as $2t$ is divisible by m or not. For odd m we have

$$|S_{a,m}|^2 = m \exp\left(2\pi i \frac{a \cdot 0}{m}\right) = m.$$

For even $m = 2m_1$ we have

$$\left| S_{a,m} \right|^2 = m \left[\exp\left(2\pi i \frac{a \cdot 0^2}{m} \right) + \exp\left(2\pi i \frac{a \cdot m_1^2}{m} \right) \right] .$$

Here the right side is equal to zero for odd m_1 and equal to $2m$ for even m_1.

γ) For any integer b we have

$$\left| S_{A,m} \right| = \left| \sum_{x=0}^{m-1} \exp\left(2\pi i \frac{Ax^2 + 2Abx}{m} \right) \right|$$

and choosing b such that $2Ab \equiv a \pmod{m}$, we again obtain the result considered in problem β).

12, a. We have

$$m \sum_{z}{}' \Phi(z) = \sum_{z} \sum_{s=M}^{M+Q-1} \sum_{a=0}^{m-1} \Phi(z) \exp\left(2\pi i \frac{a(x - z)}{m} \right) .$$

The part of the sum on the right corresponding to $a = 0$ is equal to $Q \sum_{z} \Phi(z)$; the part corresponding to the remaining values of a is numerically (problem **11, c, ch. III**)

$$< \Delta \sum_{a=1}^{m-1} \left| \sum_{s=M}^{M+Q-1} \exp\left(2\pi i \frac{-as}{m} \right) \right| < \Delta m(\ln m - \delta).$$

b, α) This follows from the theorem of problem **11, a, α)** and the theorem of problem **a.**

β) The inequality of problem α) gives $R - N = \theta \sqrt{p} \ln p$. Moreover it is evident that $R + N = Q$.

γ) It follows from the theorem of problem **11, b, β)** that the conditions of the theorem of problem **a** are satisfied if we take $m = p$, $\Phi(z) = 1$, while z runs through the values $z = x^2$; $x = 0, 1, \ldots, p - 1$. But, among the values of z there is

200

one which is congruent modulo p to 0 and two congruent modulo p with each quadratic residue of a complete residue system, and hence

$$\sum_{z'}{}' \Phi(z) = 2R, \quad \sum_z \Phi(z) = p,$$

and we obtain

$$2R = \frac{Q}{p}p + \theta\sqrt{p}\ \ln p.$$

δ) This follows from the theorem of problem **11, b, γ)** and the theorem of problem **a.**

ϵ) It follows from the theorem of problem δ) that the conditions of the theorem of problem **a** are satisfied if we set $m = p$, $\Phi(z) = 1$, while z runs through the values $z = Ax^2$; $x = M_0, M_0 + 1, \ldots, M_0 + Q_0 - 1$. Therefore

$$\sum_z{}' \Phi(z) = T, \quad \sum_z \Phi(z) = Q_0,$$

from which the required formula follows.

c. The part of the sum containing the terms with $\left(\dfrac{\alpha}{p}\right) = 1$ is equal to $p(R^2 + N^2)$, the remaining part is equal to $-2pRN$. Therefore the whole sum is equal to $p(R - N)^2$.

The part of the sum containing the terms with $a = 0$ is equal to 0. The remaining part is numerically smaller than (problem **11, c, ch. III**)

$$\sum_{a=1}^{p-1} \left| \sum_{x=M}^{M+Q-1} \exp\left(2\pi i\frac{ax}{p}\right) \right| \left| \sum_{\alpha=1}^{p-1} \left| \sum_{y=M}^{M+Q-1} \exp\left(2\pi i\frac{-a\alpha y}{p}\right) \right| \right| <$$

$$< p^2 (\ln p)^2.$$

Therefore $p(R - N)^2 < p^2 (\ln p)^2$, $R - N < \sqrt{p}\ \ln p$.

Solutions of the Problems for Chapter VI

1, a. If q is an odd prime and $a^p \equiv 1 \pmod{q}$, then a belongs to one of the exponents $\delta = 1$, p modulo q. For $\delta = 1$ we have $a \equiv 1 \pmod{q}$, for $\delta = p$ we have $q - 1 = 2px$ where x is an integer.

b. If q is an odd prime and $a^p + 1 \equiv 0 \pmod{q}$, then $a^{2p} \equiv 1 \pmod{q}$. Therefore a belongs to one of the exponents $\delta = 1, 2, p, 2p$ modulo q. The cases $\delta = 1, p$ are impossible. For $\delta = 2$ we have $a^2 \equiv 1 \pmod{q}$, $a + 1 \equiv 0 \pmod{q}$. For $\delta = 2p$ we have $q - 1 = 2px$ where x is an integer.

c. The prime divisors of $2^p - 1$ are primes of the form $2px + 1$. Let $p_1, p_2, \ldots, p_k$ be any k primes of the form $2px + 1$; the number $(p_1 p_2 \ldots p_k)^p - 1$ has a prime divisor of the form $2px + 1$ which is different from $p_1, p_2, \ldots, p_k$.

d. If q is a prime and $2^{2^n} + 1 \equiv 0 \pmod{q}$, then $2^{2^{n+1}} \equiv 1 \pmod{q}$. Therefore 2 belongs to the exponent 2^{n+1} modulo q, and hence $q - 1 = 2^{n+1}x$ where x is an integer.

2. It is evident that a belongs to the exponent n modulo $a^n - 1$. Therefore n is a divisor of $\varphi(a^n - 1)$.

3, a. Assume that we arrive at the original sequence after k operations. It is evident that the k-th operation is equivalent to the following one: consider the numbers in the sequence

$$1, 2, \ldots, n - 1, n, n, n - 1, \ldots, 2, 1, 2, \ldots$$

$$\ldots, n - 1, n, n, n - 1, \ldots, 2, 1, 2, \ldots$$

in places $1, 1 + 2^k, 1 + 2 \cdot 2^k, \ldots$. Therefore the number 2 is in the $1 + 2^k$ place. Therefore the condition considered in the problem is necessary. But it is also sufficient, since it implies that we have the following congruences modulo $2n - 1$:

$$1 \equiv 1,\ 1 + 2^k \equiv 0,\ 1 + 2 \cdot 2^k \equiv -1, \ldots$$

or

$$1 \equiv 1, \ 1 + 2^k \equiv 2, \ 1 + 2 \cdot 2^k \equiv 3, \ \ldots$$

b. The solution is analogous to the solution of problem **a.**

4. The solution of the congruence $x^\delta \equiv -1 \pmod p$ belongs to an exponent of the form $\dfrac{\delta}{\delta'}$ where δ' is a divisor of δ.

Here δ' is a multiple of d if and only if $x^{\frac{\delta}{d}} \equiv 1 \pmod p$. Considering the δ values of δ' and taking $f = 1$, we find that

$$S' = \sum_{d \setminus \delta} \mu(d) S_d, \text{ where } S' \text{ is the required number and } S_d = \frac{\delta}{d}.$$

5, a. Here (3; example c, §5) we must have $\left(\dfrac{g}{2^n + 1} \right) = -1$. This condition is satisfied for $g = 3$.

b. Here we cannot have $\left(\dfrac{g}{2p + 1} \right) = 1, \ g^2 \equiv 1 \pmod{2p + 1}$. This condition is satisfied for our values of g.

c. Here we cannot have $\left(\dfrac{g}{4p + 1} \right) = 1, \ g^4 \equiv 1 \pmod{4p + 1}$. This condition is satisfied for $g = 2$.

d. Here we cannot have $\left(\dfrac{g}{2^n p + 1} \right) = 1, \ g^{2^n} \equiv 1 \pmod{2^n p + 1}$. This condition is satisfied for $g = 3$.

6, a, α) The theorem is evident if n is a multiple of $p - 1$. Assume that n is not divisible by $p - 1$. If we disregard the order, the numbers $1, 2, \ldots, p - 1$ are congruent modulo p to the numbers $g, 2g, \ldots, (p - 1)g$, where g is a primitive root modulo p. Hence

$$S_n \equiv g^n S_n \pmod p, \ S_n \equiv 0 \pmod p.$$

β) We have

$$\sum_{x=1}^{p-1} \left(\frac{x(x^2 + 1)}{p}\right) \equiv \sum_{x=1}^{p-1} x^{\frac{p-1}{2}} (x^2 + 1)^{\frac{p-1}{2}} \pmod{p}$$

from which (problem α)) we obtain the required result.

b. For $p > 2$, we have

$$1 \cdot 2 \ldots (p - 1) \equiv g^{1+2+ \cdots +p-1} \equiv g^{\frac{p-1}{2}} \equiv -1 \pmod{p}.$$

7, a. We have $g^{\mathrm{ind}_{g_1} a} \equiv a \pmod{p}$, $\mathrm{ind}_{g_1} a \; \mathrm{ind}_g g_1 \equiv \mathrm{ind}_g a \pmod{p-1}$, $\mathrm{ind}_{g_1} a \equiv \alpha \; \mathrm{ind}_g a \pmod{p-1}$.

b. It follows from $\mathrm{ind}_g a \equiv s \pmod{n}$, $\mathrm{ind}_{g_1} a \equiv \alpha \; \mathrm{ind}_g a$ $\pmod{p-1}$ that $\mathrm{ind}_{g_1} a \equiv \alpha s \equiv s_1 \pmod{n}$.

8. Let $(n, p - 1) = 1$. Determining u by the condition $nu \equiv 1 \pmod{p-1}$ we find the solution $x \equiv a^u \pmod{p}$.

Let n be a prime, $p - 1 = n^\alpha t$, where α is a positive integer and $(t, n) = 1$. If the congruence is possible, then $a^{n^{\alpha-1} t} \equiv 1 \pmod{p}$; if $\alpha > 1$, then, noting that $x \equiv g^{n^{\alpha-1} t r}$ $\pmod{p}$, $r = 0, 1, \ldots, n - 1$ are just all the solutions of the congruence $x^n \equiv 1 \pmod{p}$; for some $r_1 = 0, 1, \ldots, n - 1$ we have

$$a^{n^{\alpha-2} t} g^{n^{\alpha-1} t} \equiv 1 \pmod{p};$$

if $\alpha > 2$, then for certain $r_2 = 0, 1, \ldots, n - 1$ we have

$$a^{n^{\alpha-3} t} g^{n^{\alpha-2} t r_1 + n^{\alpha-1} t r_2} \equiv 1 \pmod{p},$$

etc.; finally, for certain $r_{\alpha-1} = 0, 1, \ldots, n - 1$ we have

$$a^t g^{n t r_1 + n^2 t r_2 + \cdots + n^{\alpha-1} t r_{\alpha-1}} \equiv 1 \pmod{p}.$$

204

Determining u and v by the condition $tu - nv = -1$, we obtain n solutions:

$$x \equiv a^v g^{uf(r_1 + nr_2 + \ldots + n^{a-2} r_{a-1}) + n^{a-1} tr} \equiv \pmod{p};$$

$$r = 0, 1, \ldots, n - 1.$$

Let the prime n_1 divide $(n, p - 1)$, $n = n_1 n_2$, $n_2 > 1$. Corresponding to each solution of the congruence $y^{n_1} \equiv a \pmod{p}$ we obtain a solution of the congruence $x^{n_2} \equiv y \pmod{p}$.

9, a. In this way we obtain $c c_0 c_1 \ldots c_k = \varphi(m)$ characters.

b, α) We have $\chi(1) = R^0 \ldots R_k^0 = 1$.

β) Let $\gamma', \ldots, \gamma_k'; \gamma'', \ldots, \gamma_k''$ be the index systems of the numbers a_1 and a_2; then $\gamma' + \gamma'', \ldots, \gamma_k' + \gamma_k''$ is an index system for the number $a_1 a_2$ (**c,** § 7).

γ) For $a_1 \equiv a_2 \pmod{m}$, the indices of the numbers a_1 and a_2 are congruent to one-another modulo $c, \ldots, c_k$ respectively.

c. This property follows from

$$\sum_{a=0}^{m-1} \chi(a) = \sum_{\gamma=0}^{c-1} R^\gamma \ldots \sum_{\gamma_k=0}^{c_k-1} R_k^{\gamma_k}.$$

d. This property follows from

$$\chi(a) = \sum_R R^\gamma \ldots \sum_{R_k} R_k^{\gamma_k}.$$

e. Let $\psi(a_1) \gtrless 0$. Then $\psi(a_1) = \psi(a_1)\psi(1)$. Therefore $\psi(1) = 1$. Determining a' by the condition $aa' \equiv 1 \pmod{m}$, we have $\psi(a)\psi(a') = 1$. Therefore $\psi(a) \gtrless 0$ for $(a, m) = 1$.

For $(a_1, m) = 1$, we have

$$\sum_a' \frac{\chi(a)}{\psi(a)} = \sum_a \frac{\chi(a_1 a)}{\psi(a_1 a)} = \frac{\chi(a_1)}{\psi(a_1)} \sum_a' \frac{\chi(a)}{\psi(a)};$$

therefore, either $\sum_a' \dfrac{\chi(a)}{\psi(a)} = 0$, or $\psi(a_1) = \chi(a_1)$ for all a_1.

But the first cannot hold for all χ; if it did, then we would have $H = 0$, while $H = \varphi(m)$ since, summing over all characters for given a, we have

$$\sum_x \frac{\chi(a)}{\psi(a)} = \begin{cases} \varphi(m), & \text{if } a \equiv 1 \ (\text{mod } m), \\ 0, & \text{otherwise.} \end{cases}$$

f, α) If $R', \ldots, R_k'$ and $R'', \ldots, R_k''$ are values of $R, \ldots, R_k$ corresponding to the characters $\chi_1(a)$ and $\chi_2(a)$, then $\chi_1(a)\chi_2(a)$ is a character corresponding to the values $R'R'', \ldots, R_k'R_k''$.

β) When $R, \ldots, R_k$ run through all the roots of the corresponding equations, then $R'R, \ldots, R_k'R_k$ run through the same roots in some order.

γ) Determining l' by the condition $ll' \equiv 1 \ (\text{mod } m)$, we have

$$\sum_\chi \frac{\chi(a)}{\chi(l)} = \sum_\chi \frac{\chi(al')}{\chi(ll')} = \sum_\chi \chi(al')$$

which is equal to $\varphi(m)$ or 0, according as $a \equiv l \ (\text{mod } m)$ or not.

10, a, α) Defining x' by the congruence $xx' \equiv 1 \ (\text{mod } p)$, we have

$$\sum_{x=1}^{p-1} \exp\left(2\pi i \, \frac{l \text{ ind } (x + k) - l \text{ ind } x}{n}\right) =$$

$$= \sum_{x=1}^{p-1} \exp\left(2\pi i \frac{l \text{ ind } (1 + kx')}{n}\right) = -1.$$

β) We have

$$S = \sum_{x=0}^{p-1} \sum_{z_1=0}^{Q-1} \sum_{z=0}^{Q-1} \exp\left(2\pi i \frac{l \operatorname{ind}(x+z_1) - l \operatorname{ind}(x+z)}{n}\right).$$

For $z_1 = z$, summing over x gives $p-1$, and for z_1 unequal to z, summation over x (problem α)) gives -1. Therefore

$$S = (p-1)Q - Q(Q-1) = (p-Q)Q.$$

γ) Let Q_x be the number of integers of the sequence $x+z$; $z = 0, 1, \ldots, Q-1$ which are not divisible by p, while $T_{n,x}$ is the number of integers of this sequence which are in the s-th set. Finally, let

$$U_{n,x} = -\frac{Q_x}{n} + T_{n,x}, \quad S = \sum_{x=0}^{p-1} U_{n,x}^2.$$

We have

$$U_{n,x} = \frac{1}{n} \sum_{l=1}^{n-1} \sum_{z=1}^{Q-1} \exp\left(2\pi i \frac{l(\operatorname{ind}(x+z) - s)}{n}\right) =$$

$$= \frac{1}{n} \sum_{l=1}^{n-1} \exp\left(-2\pi i \frac{ls}{n}\right) S_{l,n,x},$$

$$U_{n,x}^2 \leqslant \frac{1}{n^2}(n-1) \sum_{l=1}^{n-1} \left| S_{l,n,x} \right|^2, \quad S \leqslant \left(\frac{n-1}{n}\right)^2 (p-Q)Q.$$

Setting $Q = [n\sqrt{p}\,]$, and assuming that there are no numbers of the s-th set in our sequence, we find that $|U_{n,x}| \geqslant \dfrac{Q-1}{n}$

207

for $x = M, M + 1, \ldots, M + Q + 1$, and hence

$$Q\left(\frac{Q-1}{n}\right)^2 \leqslant \left(\frac{n-1}{n}\right)^2 (p - Q)Q,$$

$$(n\sqrt{p} - 2)^2 < (n\sqrt{p} - \sqrt{p})^2,$$

which is impossible.

b. Let p_0 be the product of the different prime divisors of the number $p - 1$, let Q_x be the number of integers of the sequence $x + z$; $z = 0, 1, \ldots, Q - 1$ which are not divisible by p, and let G_x be the number of integers of the same sequence which are primitive roots modulo p. Finally, let

$$P = \left(\sum_{d\backslash p_0} \frac{\mu(d)}{d}\right)^{-1} = \frac{p-1}{\varphi(p-1)}, \quad w_x = -\frac{1}{p} + G_x,$$

$$\Omega = \sum_{x=0}^{p-1} w_x^2.$$

Taking $f(\xi) = 1$ and letting ξ run through the values $\xi = \text{ind}(x + z)$; $z = 0, 1, \ldots, Q - 1$, we obtain

$S' = \sum_{d\backslash p_0} \mu(d)S_d$. Here S' is the number of values of ξ such

that $(\xi, p - 1) = 1$ and hence $S' = G_x$. Moreover, S_d is the number of values of ξ which are multiples of d and hence $S_d = T_{d,x}$ (problem **a**, γ) for $s = 0$. Therefore

$$w_x = -\frac{1}{p} + \sum_{d\backslash p_0} \mu(d)T_{d,x} = \sum_{d\backslash p_0} \mu(d)U_{d,x},$$

$$w_x^2 \leqslant 2^k \sum_{d\backslash p_0} U_{d,x}^2, \quad \Omega \leqslant 2^k(p - Q)Q.$$

Setting $Q = [P \, 2^k \sqrt{p}\,]$ and assuming that there are no primi-
tive roots in our sequence, we find that $|w_x| \geqslant \dfrac{Q-1}{P}$ for
$x = M, M+1, \ldots, M+Q-1$ and hence

$$Q \left(\frac{Q-1}{P} \right)^2 < 2^{2k}(p-Q)Q,$$

$$(P2^k \sqrt{p} - 2)^2 < \left(P2^k \sqrt{p} - \frac{P2^k Q}{2\sqrt{p}} \right)^2,$$

which is impossible.

11, a, α). We have

$$|U_{a,p}|^2 = \sum_{t=1}^{p-1} \sum_{x=1}^{p-1} \exp\left(2\pi i \frac{k \, \text{ind} \, t}{n} \right) \exp\left(2\pi i \frac{a(t-1)x}{n} \right) =$$

$$= p - 1 - \sum_{t=2}^{p-1} \exp\left(2\pi i \frac{k \, \text{ind} \, t}{n} \right) = p.$$

β) For $(a, p) = p$ the theorem is evident. For $(a, p) = 1$,
it follows from

$$U_{a,p} =$$

$$= \exp\left(2\pi i \frac{-k \, \text{ind} \, a}{n} \right) \sum_{x=1}^{p-1} \exp\left(2\pi i \frac{k \, \text{ind} \, ax}{n} \right) \exp\left(2\pi i \frac{ax}{p} \right) =$$

$$= \exp\left(2\pi i \frac{-k \, \text{ind} \, a}{n} \right) U_{1,p}.$$

γ) It is evident that A and B are integers with $|S|^2 = A^2 + B^2$.

209

For certain ϵ, ϵ', ϵ'' such that $|\epsilon| = |\epsilon'| = |\epsilon''| = 1$ we have (problem β))

$$S = \frac{1}{\epsilon\sqrt{p}\,\epsilon\sqrt{p}} \times$$

$$\times \sum_{z_1=1}^{p-1} \sum_{z=1}^{p-1} \sum_{x=0}^{p-1} \exp\left(-2\pi i\,\frac{\text{ind }z_1 + \text{ind }z}{4}\right) \exp\left(2\pi i\,\frac{z_1 x + z(x+1)}{p}\right) .$$

If $z_1 + z$ is not equal to p, then summing over x gives zero. Therefore

$$S = \epsilon' \sum_{z=1}^{p-1} \left(\frac{z}{p}\right) \exp\left(2\pi i\,\frac{z}{p}\right) = \epsilon''\sqrt{p}\,, \quad |S|^2 = p.$$

b, α) For given z, the congruence $x^n \equiv z \pmod{p}$ is solvable only if ind z is divisible by δ, and it then has δ solutions. Therefore, for $\delta = 1$ we have $S_{a,p} = 0$. If $\delta > 1$, then we have

$$S_{a,p} = 1 + \sum_{k=0}^{\delta-1} \sum_{z=1}^{p-1} \exp\left(2\pi i\,\frac{k\,\text{ind }z}{\delta}\right) \exp\left(2\pi i\,\frac{az}{p}\right) .$$

For $k = 0$, summation with respect to z gives -1; for $k > 0$ it gives a quantity whose modulus is equal to $\sqrt{p}$. The required result follows from this.

β) Setting

$$x = u + p^{s-1}v; \; u = 0, \ldots, p^{s-1} - 1, v = 0, \ldots, p - 1,$$

we have

$$\exp\left(2\pi i\,\frac{ax^n}{p^s}\right) = \exp(2\pi i a(u^n p^{-s} + n u^{n-1} p^{-1} v)).$$

For $(u, p) = 1$, summation with respect to v gives zero.

210

Therefore

$$S_{a,p^s} = \sum_{x_0=0}^{p^{s-1}-1} \exp(2\pi i a p^{n-s} x_0^n) = p^{s-1}, \quad S'_{a,p^s} = 0.$$

γ) Let p^τ be the largest power of p dividing n. We have $s \geqslant \tau + 3$. Setting

$$x = u + p^{s-1-\tau}v; \quad u = 0, \ldots, p^{s-1-\tau} - 1, \ v = 0, \ldots, p^{\tau+1} - 1,$$

we have

$$\exp\left(2\pi i \frac{ax^n}{p^s}\right) = \exp(2\pi i a (u^n p^{-s} + n u^{n-1} p^{-\tau-1} v)).$$

For $(u, p) = 1$, summation with respect to v gives zero. Therefore

$$S_{a,p^s} = \sum_{x_0=0}^{p^{s-1}-1} \exp\left(2\pi i \frac{ax_0^n}{p^{s-n}}\right) = p^{n-1} S_{a,p^{s-n}}, \quad S'_{a,p^s} = 0.$$

δ) Let $m = p_1^{\alpha_1} \ldots p_k^{\alpha_k}$ be the canonical decomposition of the number m. Setting

$$T_{a,m} = m^{-1+\nu} S_{a,m}; \quad \nu = \frac{1}{n}, \quad m = p_1^{\alpha_1} M_1 = \ldots = p_k^{\alpha_k} M_k$$

and defining $a_1, \ldots, a_k$ by the condition $a \equiv a_1 M_1 + \ldots$
$\ldots + a_k M_k \pmod{m}$, we have (problem **12, d, ch. III**)

$$T_{a,m} = T_{a_1, p_1^{\alpha_1}} \ldots T_{a_k, p_k^{\alpha_k}}.$$

For $s = 1$ we have

$$|T_{a,p^s}| < p^{-1+\nu} n \sqrt{p} < n p^{-\frac{1}{6}}.$$

211

For $1 < s \leqslant n$, $(n, p) = 1$ we have

$$\left| T_{a,p^s} \right| = p^{-s+s} \nu p^{s-1} \leqslant 1.$$

For $1 < s \leqslant n$, $(n, p) = p$ we have

$$\left| T_{a,p^s} \right| \leqslant p^{-s+s} \nu p^s < p \leqslant n.$$

The case $s > n$ reduces to the case $s \leqslant n$ since $T_{a,p^s} = p^{-s+s} \nu p^{n-1} S_{a,p^{s-n}} = T_{a,p^{s-n}}$. Therefore

$$\left| T_{a,m} \right| \leqslant C = n^{n^6+n},$$

from which we obtain the required inequality.

12, a. This follows from the theorem of problem **11, a,** α) and the theorem of problem **12, a, ch. V.**

b. We have

$$T_n = \sum_{x=M}^{M+Q-1} \sum_{k=0}^{n-1} \exp\left(2\pi i \frac{k(\operatorname{ind} x - s)}{n}\right).$$

For $k = 0$, summing with respect to x we obtain Q; for $k > 0$, we obtain a number whose modulus is $< \sqrt{p} \ln p$. And this implies the required formula.

c. Taking $f(x) = 1$ and letting x run through the numbers $x = \operatorname{ind} M$, $\operatorname{ind} (M + 1)$, ..., $\operatorname{ind} (M + Q - 1)$, we find (problem **17, a, ch. II**) $S' = \sum_{d \backslash p-1} \mu(d) S_d$. Here S' is the number of x such that $(x, p - 1) = 1$; therefore $S' = T$. Moreover, S_d is the number of values of x which are multiples of d, i.e. the number of residues of power d in the sequence M, $M + 1$, ..., $M + Q - 1$. Then

$$H = \sum_{d \backslash p-1} \mu(d) \left(\frac{Q}{d} + \theta_d \sqrt{p} \ln p\right); \quad |\theta_d| < 1, \quad \theta_1 = 0.$$

212

d. It follows from the theorem of problem **a** that the conditions of problem **12, a, ch. V** are satisfied, if we set $m = p - 1$, $\Phi(z) = 1$, while we let z run through the values $z =$ ind x; $x = M, M + 1, \ldots, M + Q - 1$. We then find (with Q_1 in place of Q)

$$\sideset{}{'}\sum_z \Phi(z) = J, \quad \sum_z \Phi(z) = Q, \quad J = \frac{Q_1}{p - 1} Q + \theta \sqrt{p} \ (\ln p)^2.$$

13. Assume that there are no non-residues not exceeding h. The number of n-th power non-residues among the numbers $1, 2, \ldots, Q$ where

$$Q = \sqrt{p} \ (\ln p)^2$$

can be estimated by two methods: starting from the formula of problem **12, b** and starting from the fact that the non-residues can only be numbers divisible by primes exceeding h. We find

$$1 - \frac{1}{n} < \ln \frac{\dfrac{1}{2} \ln p + 2 \ln \ln p}{\dfrac{1}{c} \ln p + 2 \ln \ln p} + O\left(\frac{1}{\ln p}\right),$$

$$0 < \ln \frac{1 + 4 \dfrac{\ln \ln p}{\ln p}}{1 + 2c \dfrac{\ln \ln p}{\ln p}} + O\left(\frac{1}{\ln p}\right).$$

The impossibility of the latter inequality for all sufficiently large p proves the theorem.

14, a. We have

$$|S|^2 < X \sum_{x=0}^{m-1} \sum_{y_1=0}^{m-1} \sum_{y=0}^{m-1} \rho(y_1) \overline{\rho(y)} \exp\left(2\pi i \frac{ax(y_1 - y)}{m}\right).$$

For given y_1 and y summation with respect to x gives $Xm|\rho(y)|^2$ or zero according as $y_1 = y$ or not. Therefore

$$|S|^2 \leqslant XYm, \quad |S| \leqslant \sqrt{XYm} \ .$$

b, α) We have

$$S = \frac{1}{\varphi(m)} \sum_u \sum_v \chi(u)\chi(v) \exp\left(2\pi i \frac{au^n v^n}{m}\right)$$

where u and v run through reduced residue systems modulo m. Hence

$$S = \frac{1}{\varphi(m)} \sum_{x=0}^{m-1} \sum_{y=0}^{m-1} \nu(x)\rho(y) \exp\left(2\pi i \frac{axy}{m}\right) \ ;$$

$$\nu(x) = \sum_{u^n \equiv x(\bmod\ m)} \chi(u), \quad \rho(y) = \sum_{v^n \equiv y(\bmod\ m)} \chi(v).$$

But we have (problem **11, ch. IV**)

$$\sum_{x=0}^{m-1} |\nu(x)|^2 \leqslant K\varphi(m), \quad \sum_{y=0}^{m-1} |\rho(y)|^2 \leqslant K\varphi(m).$$

Therefore (problem **a**)

$$|S| \leqslant \frac{1}{\varphi(m)} \sqrt{K\varphi(m)\ K\varphi(m)m} = K\sqrt{m} \ .$$

β) Let $m = 2^\alpha p_1^{\alpha_1} \ldots p_k^{\alpha_k}$ be the canonical decomposition of the number m. The congruence $x^n \equiv 1 \pmod{m}$ is equivalent to the system

$$x^n \equiv 1 \pmod{2^\alpha}, \ x^n \equiv 1 \pmod{p_1^{\alpha_1}}, \ \ldots, \ x^n \equiv 1 \pmod{p_k^{\alpha_k}}.$$

214

Let $\gamma(x)$ and $\gamma_0(x)$ be the indices of the number x modulo 2^α (**g, §6**). The congruence $x^n \equiv 1 \pmod{2^\alpha}$ is equivalent to the system $n\gamma(x) \equiv 0 \pmod c$, $n\gamma_0(x) \equiv 0 \pmod{c_0}$. The first congruence of this system has at most 2 solutions, while the second has at most n solutions. Therefore the congruence $x^n \equiv 1 \pmod{2^\alpha}$ has at most $2n$ solutions. By **b, §5**, each of the congruences $x^n \equiv 1 \pmod{p_1^{\alpha_1}}$, ..., $x^n \equiv 1 \pmod{p_k^{\alpha_k}}$ have no more than n solutions. Therefore

$$K \leqslant 2n^{k+1} = 2(2^{k+1})^{\frac{\ln n}{\ln 2}} \leqslant 2(\tau(m))^{\frac{\ln n}{\ln 2}}; \quad K = 0\,(m^\epsilon).$$

15, a. We have

$$|S|^2 = \sum_{t=1}^{p-1} \sum_{x=1}^{p-1} \exp\left(2\pi i\,\frac{a(t^n - 1)x^n + b(t - 1)x}{p}\right).$$

If $t^n \equiv 1 \pmod p$, then summation with respect to x gives $p - 1$ for $t \equiv 1 \pmod p$ and -1 in the remaining cases. Otherwise, taking $z(t - 1)^{-1}$ in place of x, we can represent the part of the double sum corresponding to given t in the form

$$\sum_{z=1}^{p-1} \exp\left(2\pi i\,\frac{bz}{p}\right) \exp\left(2\pi i\,\frac{a(t^n - 1)(t - 1)^{-n}z^n}{p}\right)$$

and hence

$$|S|^2 \leqslant p - 1 + \left|\sum_{u=1}^{p-1} \sum_{v=1}^{p-1} \nu(u)\rho(v) \exp\left(2\pi i\,\frac{auv}{p}\right)\right|,$$

where $\nu(u)$ is equal to the number of solutions of the congruence $(t^n - 1)(t - 1)^{-n} \equiv u \pmod p$, while $|\rho(v)|$ does not exceed the number of solutions of the congruence $z^n \equiv v \pmod p$. Therefore $\nu(u) \leqslant 2n_1$, $|\rho(v)| < n_1$,

$$\sum_{u=1}^{p-1} |\nu(u)|^2 \leqslant (p - 1)2n_1, \quad \sum_{v=1}^{p-1} |\rho(v)|^2 \leqslant (p - 1)n_1.$$

215

Applying the theorem of problem **14, a,** we find

$$|S|^2 < p - 1 + \sqrt{(p-1)2n_1(p-1)n_1 p} < 2n_1 p^{\frac{3}{2}}.$$

b, α) This follows from problem **a** and the theorem of problem **12, a, ch. V.**

β) It follows from the theorem of problem α) that the conditions of the theorem of problem **12, a, ch. V** are satisfied, if we set $m = p$, $\Phi(z) = 1$, where we let z run through the values $z = Ax^n$; $x = M_0, M_0 + 1, \ldots, M_0 + Q_0 - 1$. Then

$$\sum_z{}' \Phi(z) = T, \quad \sum_z \Phi(z) = Q_0,$$

from which we obtain the required formula.

c, α) Let $y \equiv 4a\gamma_1 \pmod{p}$. We have (problem **11, a, ch, V**)

$$\left(\frac{a}{p}\right) S = \sum_{x=0}^{p-1} \left(\frac{4a^2x^2 + 4abx + 4ac}{p}\right) \exp\left(2\pi i \frac{4a\gamma_1 x}{p}\right) =$$

$$= \frac{1}{U_{1,p}} \sum_{z=1}^{p-1} \left(\frac{z}{p}\right) \sum_{x=0}^{p-1} \exp\left(2\pi i \frac{z(4a^2x^2 + 4abx + 4ac + 4a\gamma_1 xz^{-1})}{p}\right) =$$

$$= \sum_{z=1}^{p-1} \exp\left(2\pi i \frac{-(b^2 - 4ac)z - 2b\gamma_1 - \gamma_1^2 z^{-1}}{p}\right).$$

The latter sum (problem **a**) is numerically $< \frac{3}{2} p^{\frac{3}{4}}$.

β) This follows from the theorem of problem α) and the theorem of problem **12, a, ch. V.**

216

ANSWERS FOR THE NUMERICAL EXERCISES

Answers for chapter I.

1, **a.** 17.
   b. 23.

2, **a,** α) $\delta_4 = \dfrac{15}{11}$; β) $\alpha = \dfrac{19}{14} + \dfrac{\theta}{14 \cdot 20}$.

   b, α) $\delta_6 = \dfrac{80}{59}$; β) $\alpha = \dfrac{1002}{739} + \dfrac{\theta}{739 \cdot 1000}$.

3. We obtain 22 fractions.

5, **a.** $2^8 \cdot 3^5 \cdot 11^3$.
   b. $2^2 \cdot 3^3 \cdot 5^4 \cdot 7^3 \cdot 11^2 \cdot 17 \cdot 23 \cdot 37$.

Answers for chapter II.

1, **a.** 1312.
   b. $2^{119} \cdot 3^{59} \cdot 5^{31} \cdot 7^{19} \cdot 11^{12} \cdot 13^9 \cdot 17^7 \cdot 19^6 \cdot 23^5 \cdot 29^4 \times$
   $\times 31^4 \cdot 37^3 \cdot 41^3 \cdot 43^2 \cdot 47^2 \cdot 53^2 \cdot 59^2 \cdot 61^2 \cdot 67 \cdot 71 \times$
   $\times 73 \cdot 79 \cdot 83 \cdot 89 \cdot 97 \cdot 101 \cdot 103 \cdot 107 \cdot 109 \cdot 113$.

2, **a.** $\tau(2800) = 30$; $S(2800) = 7688$.
   b. $\tau(232\ 848) = 120$; $S(232\ 848) = 848\ 160$.

3. The sum of all the values is equal to 1.

4. α) 1152; β) 466 400.

5. The sum of all the values is equal to 774.

Answers for chapter III.

1, a. 70.
    b. It is divisible.
2, a. $3^3 \cdot 5^2 \cdot 11^2 \cdot 2999$.
    b. $7 \cdot 13 \cdot 37 \cdot 73 \cdot 101 \cdot 137 \cdot 17 \cdot 19 \cdot 257$.

Answers for chapter IV.

1, a. $x \equiv 81 \pmod{337}$.
    b. $x \equiv 200; 751; 1302; 1853; 2404 \pmod{2755}$.
2, b. $x \equiv 1630 \pmod{2413}$.
3. $x = 94 + 111t$; $y = 39 + 47t$, where t is any integer.
4, a. $x \equiv 170b_1 + 52b_2 \pmod{221}$; $x \equiv 131 \pmod{221}$;
       $x \equiv 110 \pmod{221}$; $x \equiv 89 \pmod{221}$.
    b. $x \equiv 11\ 151b_1 + 11\ 800b_2 + 16\ 875b_3 \pmod{39\ 825}$.
5, a. $x \equiv 91 \pmod{120}$.
    b. $x \equiv 8479 \pmod{15\ 015}$.
6. $x \equiv 100 \pmod{143}$; $y \equiv 111 \pmod{143}$.
7, a. $3x^4 + 2x^3 + 3x^2 + 2x \equiv 0 \pmod 5$.
    b. $x^5 + 5x^4 + 3x^2 + 3x + 2 \equiv 0 \pmod 7$.
8. $x^6 + 4x^5 + 22x^4 + 76x^3 + 70x^2 + 52x + 39 \equiv 0 \pmod{101}$.
9, a. $x \equiv 16 \pmod{27}$.
    b. $x \equiv 22; 53 \pmod{64}$.
10, a. $x \equiv 113 \pmod{125}$.
    b. $x \equiv 43, 123, 168, 248, 293, 373, 418, 498, 543, 623 \pmod{625}$.
11, a. $x \equiv 2, 5, 11, 17, 20, 26 \pmod{30}$.
    b. $x \equiv 76, 22, 176, 122 \pmod{225}$.

Answers for chapter V.

1, a. 1, 2, 3, 4, 6, 8, 9, 12, 13, 16, 18.
    b. 2, 5, 6, 8, 13, 14, 15, 17, 18, 19, 20, 22, 23, 24, 29, 31, 32, 35.
2, a. α) 0; β) 2.
    b. α) 0; β) 2.

3, a. α) 0; β) 22.

  b. α) 0; β) 2.

4, a. α) $x \equiv 9 \pmod{19}$; β) $x \equiv 11 \pmod{29}$; γ) $x \equiv 14 \pmod{97}$.

  b. α) $x \equiv 66 \pmod{311}$; β) $x \equiv 130 \pmod{277}$; γ) $x \equiv 94 \pmod{353}$.

5, a. $x \equiv 72 \pmod{125}$.

  b. $x \equiv 127 \pmod{243}$.

6, a. $x \equiv 13, 19, 45, 51 \pmod{64}$.

  b. $x \equiv 41, 87, 169, 215 \pmod{256}$.

Answers for chapter VI.

1, a. 6.

  b. 18.

2, a. 3, 3, 3.

  b. 6, 6, 1687.

  c. α) 3; β) 7.

5, a. α) 0; β) 1; γ) 3.

  b. α) 0; β) 1; γ) 10.

6, a. α) $x \equiv 40; 27 \pmod{67}$, β) $x \equiv 33 \pmod{67}$, γ) $x \equiv 8, 36, 28, 59, 31, 39 \pmod{67}$.

  b. α) $x \equiv 17 \pmod{73}$; β) $x \equiv 50, 12, 35, 23, 61, 38 \pmod{73}$, γ) $x \equiv 3, 24, 46 \pmod{73}$.

7, a. α) 0; β) 4.

  b. α) 0; β) 7.

8, a. α) $x \equiv 54 \pmod{101}$. β) $x \equiv 53, 86, 90, 66, 8 \pmod{101}$.

  b. $x \equiv 59, 11, 39 \pmod{109}$.

9, a. α) 1, 4, 5, 6, 7, 9, 11, 16, 17; β) 1, 7, 8, 11, 12, 18.

  b. α) 1, 6, 8, 10, 11, 14, 23, 26, 27, 29, 31, 36; β) 1, 7, 9, 10, 12, 16, 26, 33, 34.

10, a. α) 7, 37; β) 3, 5, 12, 18, 19, 20, 26, 28, 29, 30, 33, 34.

  b. α) 3, 27, 41, 52; β) 2, 6, 7, 10, 17, 18, 23, 26, 30, 31, 35, 43, 44, 51, 54, 55, 59.

TABLES OF INDICES

The Prime 3

N	0	1	2	3	4	5	6	7	8	9
0		0	1							

I	0	1	2	3	4	5	6	7	8	9
0	1	2								

The Prime 5

N	0	1	2	3	4	5	6	7	8	9
0		0	1	3	2					

I	0	1	2	3	4	5	6	7	8	9
0	1	2	4	3						

The Prime 7

N	0	1	2	3	4	5	6	7	8	9
0		0	2	1	4	5	3			

I	0	1	2	3	4	5	6	7	8	9
0	1	3	2	6	4	5				

The Prime 11

N	0	1	2	3	4	5	6	7	8	9
0		0	1	8	2	4	9	7	3	6
1	5									

I	0	1	2	3	4	5	6	7	8	9
0	1	2	4	8	5	10	9	7	3	6
1										

The Prime 13

N	0	1	2	3	4	5	6	7	8	9
0		0	1	4	2	9	5	11	3	8
1	10	7	6							

I	0	1	2	3	4	5	6	7	8	9
0	1	2	4	8	3	6	12	11	9	5
1	10	7								

The Prime 17

N	0	1	2	3	4	5	6	7	8	9
0		0	14	1	12	5	15	11	10	2
1	3	7	13	4	9	6	8			

I	0	1	2	3	4	5	6	7	8	9
0	1	3	9	10	13	5	15	11	16	14
1	8	7	4	12	2	6				

The Prime 19

N	0	1	2	3	4	5	6	7	8	9
0		0	1	13	2	16	14	6	3	8
1	17	12	15	5	7	11	4	10	9	

I	0	1	2	3	4	5	6	7	8	9
0	1	2	4	8	16	13	7	14	9	18
1	17	15	11	3	6	12	5	10		

The Prime 23

N	0	1	2	3	4	5	6	7	8	9
0		0	2	16	4	1	18	19	6	10
1	3	9	20	14	21	17	8	7	12	15
2	5	13	11							

I	0	1	2	3	4	5	6	7	8	9
0		1	5	2	10	4	20	8	17	16	11
1		9	22	18	21	13	19	3	15	6	7
2	12	14								

The Prime 29

N	0	1	2	3	4	5	6	7	8	9
0		0	1	5	2	22	6	12	3	10
1	23	25	7	18	13	27	4	21	11	9
2	24	17	26	20	8	16	19	15	14	

I	0	1	2	3	4	5	6	7	8	9
0	1	2	4	8	16	3	6	12	24	19
1	9	18	7	14	28	27	25	21	13	26
2	23	17	5	10	20	11	22	15		

The Prime 31

N	0	1	2	3	4	5	6	7	8	9
0		0	24	1	18	20	25	28	12	2
1	14	23	19	11	22	21	6	7	26	4
2	8	29	17	27	13	10	5	3	16	9
3	15									

I	0	1	2	3	4	5	6	7	8	9
0	1	3	9	27	19	26	16	17	20	29
1	25	13	8	24	10	30	28	22	4	12
2	5	15	14	11	2	6	18	23	7	21

The Prime 37

N	0	1	2	3	4	5	6	7	8	9
0		0	1	26	2	23	27	32	3	16
1	24	30	28	11	33	13	4	7	17	35
2	25	22	31	15	29	10	12	6	34	21
3	14	9	5	20	8	19	18			

I	0	1	2	3	4	5	6	7	8	9
0	1	2	4	8	16	32	27	17	34	31
1	25	13	26	15	30	23	9	18	36	35
2	33	29	21	5	10	20	3	6	12	24
3	11	22	7	14	28	19				

The Prime 41

N	0	1	2	3	4	5	6	7	8	9
0		0	26	15	12	22	1	39	38	30
1	8	3	27	31	25	37	24	33	16	9
2	34	14	29	36	13	4	17	5	11	7
3	23	28	10	18	19	21	2	32	35	6
4	20									

I	0	1	2	3	4	5	6	7	8	9
0	1	6	36	11	25	27	39	29	10	19
1	32	28	4	24	21	3	18	26	33	34
2	40	35	5	30	16	14	2	12	31	22
3	9	13	37	17	20	38	23	15	8	7

The Prime 43

N	0	1	2	3	4	5	6	7	8	9
0		0	27	1	12	25	28	35	39	2
1	10	30	13	32	20	26	24	38	29	19
2	37	36	15	16	40	8	17	3	5	41
3	11	34	9	31	23	18	14	7	4	33
4	22	6	21							

I	0	1	2	3	4	5	6	7	8	9
0	1	3	9	27	38	28	41	37	25	32
1	10	30	4	12	36	22	23	26	35	19
2	14	42	40	34	16	5	15	2	6	18
3	11	33	13	39	31	7	21	20	17	8
4	24	29								

The Prime 47

N	0	1	2	3	4	5	6	7	8	9
0		0	18	20	36	1	38	32	8	40
1	19	7	10	11	4	21	26	16	12	45
2	37	6	25	5	28	2	29	14	22	35
3	39	3	44	27	34	33	30	42	17	31
4	9	15	24	13	43	41	23			

I	0	1	2	3	4	5	6	7	8	9
0	1	5	25	31	14	23	21	11	8	40
1	12	13	18	43	27	41	17	38	2	10
2	3	15	28	46	42	22	16	33	24	26
3	36	39	7	35	34	29	4	20	6	30
4	9	45	37	44	32	19				

The Prime 53

N	0	1	2	3	4	5	6	7	8	9
0		0	1	17	2	47	18	14	3	34
1	48	6	19	24	15	12	4	10	35	37
2	49	31	7	39	20	42	25	51	16	46
3	13	33	5	23	11	9	36	30	38	41
4	50	45	32	22	8	29	40	44	21	28
5	43	27	26							

I	0	1	2	3	4	5	6	7	8	9
0	1	2	4	8	16	32	11	22	44	35
1	17	34	15	30	7	14	28	3	6	12
2	24	48	43	33	13	26	52	51	49	45
3	37	21	42	31	9	18	36	19	38	23
4	46	39	25	50	47	41	29	5	10	20
5	40	27								

The Prime 59

N	0	1	2	3	4	5	6	7	8	9
0		0	1	50	2	6	51	18	3	42
1	7	25	52	45	19	56	4	40	43	38
2	8	10	26	15	53	12	46	34	20	28
3	57	49	5	17	41	24	44	55	39	37
4	9	14	11	33	27	48	16	23	54	36
5	13	32	47	22	35	31	21	30	29	

I	0	1	2	3	4	5	6	7	8	9
0	1	2	4	8	16	32	5	10	20	40
1	21	42	25	50	41	23	46	33	7	14
2	28	56	53	47	35	11	22	44	29	58
3	57	55	51	43	27	54	49	39	19	38
4	17	34	9	18	36	13	26	52	45	31
5	3	6	12	24	48	37	15	30		

The Prime 61

N	0	1	2	3	4	5	6	7	8	9
0		0	1	6	2	22	7	49	3	12
1	23	15	8	40	50	28	4	47	13	26
2	24	55	16	57	9	44	41	18	51	35
3	29	59	5	21	48	11	14	39	27	46
4	25	54	56	43	17	34	58	20	10	38
5	45	53	42	33	19	37	52	32	36	31
6	30									

I	0	1	2	3	4	5	6	7	8	9
0	1	2	4	8	16	32	3	6	12	24
1	48	35	9	18	36	11	22	44	27	54
2	47	33	5	10	20	40	19	38	15	30
3	60	59	57	53	45	29	58	55	49	37
4	13	26	52	43	25	50	39	17	34	7
5	14	28	56	51	41	21	42	23	46	31

The Prime 67

N	0	1	2	3	4	5	6	7	8	9
0		0	1	39	2	15	40	23	3	12
1	16	59	41	19	24	54	4	64	13	10
2	17	62	60	28	42	30	20	51	25	44
3	55	47	5	32	65	38	14	22	11	58
4	18	53	63	9	61	27	29	50	43	46
5	31	37	21	57	52	8	26	49	45	36
6	56	7	48	35	6	34	33			

I	0	1	2	3	4	5	6	7	8	9
0	1	2	4	8	16	32	64	61	55	43
1	19	38	9	18	36	5	10	20	40	13
2	26	52	37	7	14	28	56	45	23	46
3	25	50	33	66	65	63	59	51	35	3
4	6	12	24	48	29	58	49	31	62	57
5	47	27	54	41	15	30	60	53	39	11
6	22	44	21	42	17	34				

The Prime 71

N	0	1	2	3	4	5	6	7	8	9
0		0	6	26	12	28	32	1	18	52
1	34	31	38	39	7	54	24	49	58	16
2	40	27	37	15	44	56	45	8	13	68
3	60	11	30	57	55	29	64	20	22	65
4	46	25	33	48	43	10	21	9	50	2
5	62	5	51	23	14	59	19	43	4	3
6	66	69	17	53	36	67	63	47	61	41
7	35									

I	0	1	2	3	4	5	6	7	8	9
0	1	7	49	59	58	51	2	14	27	47
1	45	31	4	28	54	23	19	62	8	56
2	37	46	38	53	16	41	3	21	5	35
3	32	11	6	42	10	70	64	22	12	13
4	20	69	57	44	24	26	40	67	43	17
5	48	52	9	63	15	34	25	33	18	55
6	30	68	50	66	36	39	60	65	29	61

The Prime 73

N	0	1	2	3	4	5	6	7	8	9
0		0	8	6	16	1	14	33	24	12
1	9	55	22	59	41	7	32	21	20	62
2	17	39	63	46	30	2	67	18	49	35
3	15	11	40	61	29	34	28	64	70	65
4	25	4	47	51	71	13	54	31	38	66
5	10	27	3	53	26	56	57	68	43	5
6	23	58	19	45	48	60	69	50	37	52
7	42	44	36							

I	0	1	2	3	4	5	6	7	8	9
0	1	5	25	52	41	59	3	15	2	10
1	50	31	9	45	6	30	4	20	27	62
2	18	17	12	60	8	40	54	51	36	34
3	24	47	16	7	35	29	72	68	48	21
4	32	14	70	58	71	63	23	42	64	28
5	67	43	69	53	46	11	55	56	61	13
6	65	33	19	22	37	39	49	26	57	66
7	38	44								

The Prime 79

N	0	1	2	3	4	5	6	7	8	9
0		0	4	1	8	62	5	53	12	2
1	66	68	9	34	57	63	16	21	6	32
2	70	54	72	26	13	46	38	3	61	11
3	67	56	20	69	25	37	10	19	36	35
4	74	75	58	49	76	64	30	59	17	28
5	50	22	42	77	7	52	65	33	15	31
6	71	45	60	55	24	18	73	48	29	27
7	41	51	14	44	23	47	40	43	39	

I	0	1	2	3	4	5	6	7	8	9
0	1	3	9	27	2	6	18	54	4	12
1	36	29	8	24	72	58	16	48	65	37
2	32	17	51	74	64	34	23	69	49	68
3	46	59	19	57	13	39	38	35	26	78
4	76	70	52	77	73	61	25	75	67	43
5	50	71	55	7	21	63	31	14	42	47
6	62	28	5	15	45	56	10	30	11	33
7	20	60	22	66	40	41	44	53		

The Prime 83

N	0	1	2	3	4	5	6	7	8	9
0		0	1	72	2	27	73	8	3	62
1	28	24	74	77	9	17	4	56	63	47
2	29	80	25	60	75	54	78	52	10	12
3	18	38	5	14	57	35	64	20	48	67
4	30	40	81	71	26	7	61	23	76	16
5	55	46	79	59	53	51	11	37	13	34
6	19	66	39	70	6	22	15	45	58	50
7	36	33	65	69	21	44	49	32	68	43
8	31	42	41							

I	0	1	2	3	4	5	6	7	8	9
0	1	2	4	8	16	32	64	45	7	14
1	28	56	29	58	33	66	49	15	30	60
2	37	74	65	47	11	22	44	5	10	20
3	40	80	77	71	59	35	70	57	31	62
4	41	82	81	79	75	67	51	19	38	76
5	69	55	27	54	25	50	17	34	68	53
6	23	46	9	18	36	72	61	39	78	73
7	63	43	3	6	12	24	48	13	26	52
8	21	42								

The Prime 89

N	0	1	2	3	4	5	6	7	8	9
0		0	16	1	32	70	17	81	48	2
1	86	84	33	23	9	71	64	6	18	35
2	14	82	12	57	49	52	39	3	25	59
3	87	31	80	85	22	63	34	11	51	24
4	30	21	10	29	28	72	73	54	65	74
5	68	7	55	78	19	66	41	36	75	43
6	15	69	47	83	8	5	13	56	38	58
7	79	62	50	20	27	53	67	77	40	42
8	46	4	37	61	26	76	45	60	44	

I	0	1	2	3	4	5	6	7	8	9
0	1	3	9	27	81	65	17	51	64	14
1	42	37	22	66	20	60	2	6	18	54
2	73	41	34	13	39	28	84	74	44	43
3	40	31	4	12	36	19	57	82	68	26
4	78	56	79	59	88	86	80	62	8	24
5	72	38	25	75	47	52	67	23	69	29
6	87	83	71	35	16	48	55	76	50	61
7	5	15	45	46	49	58	85	77	53	70
8	32	7	21	63	11	33	10	30		

The Prime 97

N	0	1	2	3	4	5	6	7	8	9
0		0	34	70	68	1	8	31	6	44
1	35	6	42	25	65	71	40	89	78	81
2	69	5	24	77	76	2	59	18	3	13
3	9	46	74	60	27	32	16	91	19	95
4	7	85	39	4	58	45	15	84	14	62
5	36	63	93	10	52	87	37	55	47	67
6	43	64	80	75	12	26	94	57	61	51
7	66	11	50	28	29	72	53	21	33	30
8	41	88	23	17	73	90	38	83	92	54
9	79	56	49	20	22	82	48			

I	0	1	2	3	4	5	6	7	8	9
0	1	5	25	28	43	21	8	40	6	30
1	53	71	64	29	48	46	36	83	27	38
2	93	77	94	82	22	13	65	34	73	74
3	79	7	35	78	2	10	50	56	86	42
4	16	80	12	60	9	45	31	58	96	92
5	72	69	54	76	89	57	91	67	44	26
6	33	68	49	51	61	14	70	59	4	20
7	3	15	75	84	32	63	24	23	18	90
8	62	19	95	87	47	41	11	55	81	17
9	85	37	88	52	66	39				

Table of primes < 4000 and their smallest primitive roots.

p	g	p	g	p	g	p	g	p	g	p	g	p	g
2	1	179	2	419	2	661	2	947	2	1 229	2	1 523	2
3	2	181	2	421	2	673	5	953	3	1 231	3	1 531	2
5	2	191	19	431	7	677	2	967	5	1 237	2	1 543	5
7	3	193	5	433	5	683	5	971	6	1 249	7	1 549	2
11	2	197	2	439	15	691	3	977	3	1 259	2	1 553	3
13	2	199	3	443	2	701	2	983	5	1 277	2	1 559	19
17	3	211	2	449	3	709	2	991	6	1 279	3	1 567	3
19	2	223	3	457	13	719	11	997	7	1 283	2	1 571	2
23	5	227	2	461	2	727	5	1 009	11	1 289	6	1 579	3
29	2	229	6	463	3	733	6	1 013	3	1 291	2	1 583	5
31	3	233	3	467	2	739	3	1 019	2	1 297	10	1 597	11
37	2	239	7	479	13	743	5	1 021	10	1 301	2	1 601	3
41	6	241	7	487	3	751	3	1 031	14	1 303	6	1 607	5
43	3	251	6	491	2	757	2	1 033	5	1 307	2	1 609	7
47	5	257	3	499	7	761	6	1 039	3	1 319	13	1 613	3
53	2	263	5	503	5	769	11	1 049	3	1 321	13	1 619	2
59	2	269	2	509	2	773	2	1 051	7	1 327	3	1 621	2
61	2	271	6	521	3	787	2	1 061	2	1 361	3	1 627	3
67	2	277	5	523	2	797	2	1 063	3	1 367	5	1 637	2
71	7	281	3	541	2	809	3	1 069	6	1 373	2	1 657	11
73	5	283	3	547	2	811	3	1 087	3	1 381	2	1 663	3
79	3	293	2	557	2	821	2	1 091	2	1 399	13	1 667	2
83	2	307	5	563	2	823	3	1 093	5	1 409	3	1 669	2
89	3	311	17	569	3	827	2	1 097	3	1 423	3	1 693	2
97	5	313	10	571	3	829	2	1 103	5	1 427	2	1 697	3
101	2	317	2	577	5	839	11	1 109	2	1 429	6	1 699	3
103	5	331	3	587	2	853	2	1 117	2	1 433	3	1 709	3
107	2	337	10	593	3	857	3	1 123	2	1 439	7	1 721	3
109	6	347	2	599	7	859	2	1 129	11	1 447	3	1 723	3
113	3	349	2	601	7	863	5	1 151	17	1 451	2	1 733	2
127	3	353	3	607	3	877	2	1 153	5	1 453	2	1 741	2
131	2	359	7	613	2	881	3	1 163	5	1 459	5	1 747	2
137	3	367	6	617	3	883	2	1 171	2	1 471	6	1 753	7
139	2	373	2	619	2	887	5	1 181	7	1 481	3	1 759	6
149	2	379	2	631	3	907	2	1 187	2	1 483	2	1 777	5
151	6	383	5	641	3	911	17	1 193	3	1 487	5	1 783	10
157	5	389	2	643	11	919	7	1 201	11	1 489	14	1 787	2
163	2	397	5	647	5	929	3	1 213	2	1 493	2	1 789	6
167	5	401	3	653	2	937	5	1 217	3	1 499	2	1 801	11
173	2	409	21	659	2	941	2	1 223	5	1 511	11	1 811	6

(*continued*)

p	g	p	g	p	g	p	g	p	g	p	g	p	g
1 823	5	2 129	3	2 417	3	2 729	3	3 049	11	3 373	5	3 691	2
1 831	3	2 131	2	2 423	5	2 731	3	3 061	6	3 389	3	3 697	5
1 847	5	2 137	10	2 437	2	2 741	2	3 067	2	3 391	3	3 701	2
1 861	2	2 141	2	2 441	6	2 749	6	3 079	6	3 407	5	3 709	2
1 867	2	2 143	3	2 447	5	2 753	3	3 083	2	3 413	2	3 719	7
1 871	14	2 153	3	2 459	2	2 767	3	3 089	3	3 433	5	3 727	3
1 873	10	2 161	23	2 467	2	2 777	3	3 109	6	3 449	3	3 733	2
1 877	2	2 179	7	2 473	5	2 789	2	3 119	7	3 457	7	3 739	7
1 879	6	2 203	5	2 477	2	2 791	6	3 121	7	3 461	2	3 761	3
1 889	3	2 207	5	2 503	3	2 797	2	3 137	3	3 463	3	3 767	5
1 901	2	2 213	2	2 521	17	2 801	3	3 163	3	3 467	2	3 769	7
1 907	2	2 221	2	2 531	2	2 803	2	3 167	5	3 469	2	3 779	2
1 913	2	2 237	2	2 539	2	2 819	2	3 169	7	3 491	2	3 793	5
1 931	2	2 239	3	2 543	5	2 833	5	3 181	7	3 499	2	3 797	2
1 933	5	2 243	2	2 549	2	2 837	2	3 187	2	3 511	7	3 803	2
1 949	2	2 251	7	2 551	6	2 843	2	3 191	11	3 517	2	3 821	3
1 951	3	2 267	2	2 557	2	2 851	2	3 203	2	3 527	5	3 823	3
1 973	2	2 269	2	2 579	2	2 857	11	3 209	3	3 529	17	3 833	3
1 979	2	2 273	3	2 591	2	2 861	2	3 217	5	3 533	2	3 847	5
1 987	2	2 281	7	2 593	7	2 879	7	3 221	10	3 539	2	3 851	2
1 993	5	2 287	19	2 609	3	2 887	5	3 229	6	3 541	7	3 853	2
1 997	2	2 293	2	2 617	5	2 897	3	3 251	6	3 547	2	3 863	5
1 999	3	2 297	5	2 621	2	2 903	2	3 253	2	3 557	2	3 877	2
2 003	5	2 309	2	2 633	3	2 909	2	3 257	3	3 559	3	3 881	13
2 011	3	2 311	3	2 647	3	2 917	5	3 259	3	3 571	2	3 889	11
2 017	5	2 333	2	2 657	3	2 927	5	3 271	3	3 581	2	3 907	2
2 027	2	2 339	2	2 659	2	2 939	2	3 299	2	3 583	3	3 911	13
2 029	2	2 341	7	2 663	5	2 953	13	3 301	6	3 593	3	3 917	2
2 039	7	2 347	3	2 671	7	2 957	2	3 307	2	3 607	5	3 919	3
2 053	2	2 351	13	2 677	2	2 963	2	3 313	10	3 613	2	3 923	2
2 063	5	2 357	2	2 683	2	2 969	3	3 319	6	3 617	3	3 929	3
2 069	2	2 371	2	2 687	5	2 971	10	3 323	2	3 623	5	3 931	2
2 081	3	2 377	5	2 689	19	2 999	17	3 329	3	3 631	21	3 943	3
2 083	2	2 381	3	2 693	2	3 001	14	3 331	3	3 637	2	3 947	2
2 087	5	2 383	5	2 699	2	3 011	2	3 343	5	3 643	2	3 967	6
2 089	7	2 389	2	2 707	2	3 019	2	3 347	2	3 659	2	3 989	2
2 099	2	2 393	3	2 711	7	3 023	5	3 359	11	3 671	13		
2 111	7	2 399	11	2 713	5	3 037	2	3 361	22	3 673	5		
2 113	5	2 411	6	2 719	3	3 041	3	3 371	2	3 677	2		